American Brahman,

A History
of
The American Brahman

And God made the beast of the earth after his kind, and cattle after their kind, and everything that creepeth upon the earth after his kind: and God saw that it was good.

Genesis 1:25

May. 1997

To:
Ms. Muske
It's good to know
Good people like you
-Charles

American Brahman

A History of The American Brahman

Joe A. Akerman, Jr.
North Florida Junior College

AMERICAN BRAHMAN BREEDERS ASSOCIATION
1313 LA CONCHA LANE HOUSTON, TEXAS 77054

Art work by Jules de R. Bacot
North Florida Junior College

printed for the publishers by

jimbob printing, inc.
madison, florida

To my wife
Princess

Preface

"To every thing there is a season, and a time to every purpose under the heaven."

Ecclesiastes 3:1

The ownership and husbandry of cattle is as old as civilization itself, but nowhere has cattle raising been more of a social or economic agent in determining the character and occupation of its people than in the United States.

There was a time in the late 1870's and 1880's when a man on horseback could ride from the Rio Grande to the Canadian borders, from the Mississippi to the Rockies and seldom be out of sight of cattle or ever see a fence. It was a world teaming with life, a veritable sea of cattle with few communities of over 5,000 people. It was a time when waters ran clear and native grasses waved a vast carpet across the horizons. It was a time that left a great legend, the epic of the cowboy, the outlaw and the gambler. But it was also a time short-lived, a time when lands became overgrazed, when entire cattle herds were destroyed by diseases, droughts, and bitter cold weather, a time when homesteaders challenged the cattle lords for the use of public lands.

Above all, it was a time that saw the need for a more progressive, innovative and realistic approach to cattle raising and the use of land. It was a time that called for a new cattle breed, a breed that could prevail and even prosper with less supplementary feeding, that could better withstand virulent livestock diseases and insects and one that could be successfully crossed with existing native stock, producing more adaptible breeds. It was a time for developing a breed that would one day be known as the American Brahman.

It was also a time that called for a new kind of man, a man aware and appreciative of nature's precarious balance and one willing to come to grips with its limitations. It was a time for a man committed to developing a better animal - one that could help better to meet the needs of a burgeoning nation and a hungry world.

This story is only a small part of the saga dealing with this *new time*, this *new breed* and this *new man*. Because of the limitations imposed by space, many of the main characters, both man and beast, are not discussed. For this I offer my sincere apologies. I only hope that this brief history will encourage others to investigate this subject area further.

Joe A. Akerman, Jr.
North Florida Junior College
Madison, Florida

Abel P. Borden

Foreward

"When love and skill work together, expect a masterpiece."
John Ruskin

The lengthy journey of cattle from Europe and Asia to the New World and their adaptation therein is the continuing, fascinating saga of man and his cattle across the centuries . . . from the Biblical "cattle-on-a-thousand-hills" to the complex twentieth century world beef cattle industry. Of singular economic, political and social importance within the history of that journey is the moulding of an old creature made new . . . the American Brahman. And, of equal importance, the dauntless men and women who developed it. Indeed, with "love and skill" and work, they produced a masterpiece of the bovine world which has contributed mightily to the economic maturity of Mother Earth's animal agriculture and the well-being of her peoples. In less than a century, the American Brahman as a unique genetic reservoir distributed in the United States and sixty other countries has truly become "Crossbreeding's Common Denominator". As such the breed is generally utilized to compliment rather than compete with other breeds and is daily writing new chapters in the annals of successful beef production wherever used.

Here follows author and able historian Joe A. Akerman's captivating chronicle of those stalwart men and women who early perceived the majesty and economic worth of "those strange humped cattle of India" and their succeeding generations of dedicated, skilled seed-stock producers who today constitute the American Brahman Breeders Assocation. This is their story . . . . their pastures, their cattle, their heartbreaks, their ultimate victories. It could only happen in America, within its free enterprise system with the right to dare, to risk, to win. And their eternal quest for that "perfect bull or cow" continues.

ABBA Historical Data Committee

Joseph B. Dial
Harry P. Gayden
Edgar H. Hudgins
Wendell Schronk
Margaret Watkins
J. D. Sartwelle, Chairman

Acknowledgements

Many kind people have aided me in the task of researching, compiling, organizing and bringing the *American Brahman* to its present stage. I am especially grateful for the assistance and encouragement given to me by my professional colleagues at North Florida Junior College and to the NFJC Board of Trustees who granted me a sabbatical leave to complete this book. The following persons in particular should be thanked: Dr. Gary Sims, NFJC President, Mrs. Lou Sands, Director of Library Services, Mr. Jules de R. Bacot, NFJC Professor of Art and Mrs. Fanny Walton, NFJC Professor of English.

Without the guidance, direction and encouragement given to me by the Historical Data Committee of the American Brahman Breeders Association, this book would not have been completed. Chaired by J. D. Sartwelle, Sr., former ABBA Director, other members of this committee included Margaret Watkins, ABBA Honorary Vice-president and former ABBA Recording Secretary; Harry Gayden, ABBA Honorary Vice-president and ABBA Executive Secretary Emeritus; Joe Dial, ABBA Director; Edgar Hudgins, ABBA Director and former ABBA President and Wendell Schronk, ABBA Executive Vice-president. During the past three years, this committee has also supplied me with a store of documented materials.

Special thanks should also be extended to Sheila Schronk. She was an invaluable aid in helping to proofread the final manuscript.

I would like to thank all the members of the ABBA staff, especially Dinah Weil, Lee Pritchard, Charles Greenwood and Cindy Hejl.

To Mrs. Allen Morris of the State Photographic Archives for the State of Florida, special mention should be made for supplying me with several historical photographs that appear in this book.

I would also like to thank the following persons and organizations for providing me with photographs, family papers, private collections, and unpublished and published manuscripts: John Abbot of Harlingen, Texas; J. D. Sartwelle, Jr. of Houston, Texas; the late Oscar "Slim" L. Partin of Kissimmee, Florida; Jim Nettles and the Diamond T Ranch of Bay City, Texas; Joe Ed Brockett of *The Brahman Journal;* Mr. and Mrs. Robert L. Massey of Inez, Texas; Avon Davis of Chowchilla, California; John E. Rouse of Saratoga, Wyoming; The University of Oklahoma Press, Norman, Oklahoma; Gerry T. Stack of Tampa, Florida; Charles Lykes of Tampa, Florida; Vernon Frost of Simonton, Texas; J. Lewis Patterson, Concord, North Carolina; Mr. and Mrs. J. T. White of Hearne, Texas; Sloan and Jim Williams of Hungerford, Texas; Steve Allen of Madisonville, Texas; Mrs. Margie Rogillio of Greenwell Springs, Louisiana; Tom Dean of Cherokee, Texas; Henry Clay Koontz of Placedo, Texas; Emily Dial of Placedo, Texas; Byron Johnson of Ocala, Florida; Duffy G. Marceaux of Kaplan, Louisiana; M. E. "Max" Hammond of Bartow, Florida; Monty Banks of Mt. Pleasant, Texas; Herman Beville of Bushnell, Florida; Dr. John Stiles of the Florida Department of Agriculture and Consumer Services; Ron and Martha Smoak of Madison, Florida; Bill and Leah Kinner of Madison, Florida; James Coker of Hayneville, Alabama; W. C. "Cal" Burns, Beef and Cattle Research Center, Brooksville, Florida; Greg Carlton, U.S. Sugar Corporation of Clewiston, Florida; Claude Dobbins of C. Clausen Steamship Co., Alexandria, Virginia; Marvin Gamble of Perry, Florida; Dr. Marvin Koger of the University of Florida; Dr. W. C. McCormick, Coastal Plain Experiment Station, Tifton, Georgia; Roy Martin, *The Zebu Journal,* Cotulla, Texas; Dr. James Pace (ret.), University of Florida; Billy Peeples of Moore Haven, Florida; Dave Pingrey of the American Angus Association; Rick Ramsey of Houston, Texas; Dr. James O. Sanders, Texas A&M; Dr. E. C. "Buck" Sharman, USDA, Hyattsville, Maryland; Gilbert and Andy Tucker of Cocoa, Florida; Dr. William Turner, Louisiana State University; Dr. Frank D. Y'Turria of Brownsville, Texas; J. V. Gates, Poteet, Texas; Lt. Col. Roger McKinley John of the U.S. Army, New York, New York; Lloyd Clyburn, former ABBA Director of Information and Education; Kirby Cunningham, former Executive Secretary of the American Brahman Breeders Association, and Sloan and Jim Williams of Hungerford, Texas.

To my wife Princess and to my children Mark and Amy go thanks for their patience and encouragement while this book was being completed. A special thanks goes to Princess for typing the final manuscript.

I am also deeply indebted to Jim and Sylvia Williams of Jim-Bob Printing, Inc. of Madison, Florida who did so many extra things in bringing about the publication of this book.

Lastly, I would like to thank the many Brahman breeders, past and present, who lived the history this book attempts to record.

Contents

1
What is the American Brahman?

The modern American Brahman is an American achievement, the product of over a century of selective breeding and genetic improvement. But like most American creations its deepest roots are Old World. Its principal progenitors were from the great subcontinent of India where cattle have been domesticated and venerated for thousands of years.[1]

Indian cattle are scientifically classified as *Bos indicus*, but there are over thirty distinct breeds which have been customarily named for the geographical areas in which they are found. Size and conformation are as diverse as the wide variety of habitats from which they originate, and they vary from massive beasts of burden and long-legged trotters to the small Red Sindhi cattle of the Indian hills.[2]

What has made the American Brahman such a superior and special breed is that it was developed from an amalgamation of four predominant Indian breeds bred with early native range stock under American conditions with a particular beef standard in mind.[3]

While the principal bloodlines of the American Brahman were Indian, its birthplace and its phylogeny have been strictly American. It was primarily perfected on the Southern Gulf

The ability to utilize available feed, to travel necessary distances for feed and water, to resist insects and diseases and to withstand climatic extremes have characterized the American Brahman for many years, and have made them popular over much of the world.

Coastal Plains -- more specifically between the great Mississippi and Rio Grande Rivers.

It was here that a number of visionary cattlemen felt challenged and driven to perfect a new bovine species that would thrive in a region and climate where insects, disease and temperature extremes were common -- in an environment where cattle needed excellent rustling ability and the capacity to cover vast stretches of rough terrain. But they also wanted a breed that had the

The amalgamation of four principal breeds of *Bos indicus* **cattle and "native American range cattle" under American conditions has made the American Brahman a superior animal. Progeny from such herd sires are found in nearly 70 different countries and in most of the U.S.**

beef qualities of the British breeds and the adaptability of the Spanish scrubs.[4]

It was a cattleman's dream, and the odds against creating such a complete animal were monumental. Some scoffed at the idea -- after all it usually took centuries of experimentation, environmental conditioning and careful selection to perfect a new breed. How could these grotesque creatures from Asia with their floppy ears and loose, slick coats and buffalo like backs ever become Americanized? But it would not be the first, nor the last, American dream to be fulfilled.

Perhaps it was only fortuitous, but the four Indian breeds that became so important in developing the American Brahman could not have been better selected. First of all, there was the Guzerat (or Kankrej). It was as old as civilization itself, a beast of burden and an object of veneration for milleniums. Among the Hindus of India they represented the pastoral life which they considered idyllic.[5] Pictures and carvings taken from archealogical excavations in Pakistan show that cattle of the Guzerat type existed as far back as 3,000 B.C.[6]

The home of this prized breed is the strategic Rann of Cutch, which straddles the west end of the Indian and West Pakistanian border, but they may be found in an area that extends from here for about 700 miles. Its breeding track in India and Pakistan is generally low-lying and dry and the country is a sandy, treeless plain. The soil types vary considerably from rich heavy black to sandy or low nutrient clay. Temperatures also range markedly from 40° F to 120°F, and like most of India, the precipitation is very seasonal. Because of such a disparity of climatic and range conditions, the Guzerat has been forced to make many accommodations in its long and fascinating history. The attributes acquired from these extremes of conditioning explain, in part, why the American Brahman became such a self-sufficient and versatile breed.[7]

From 1835 to 1906, the Guzerat was the Indian breed most often imported into the United States and South America. Distinguished by its large horns, its broad chest, its straight back and its powerful appearance, it was originally used in the New World as a draft animal on the great haciendas and plantations. However, early New World cattle raisers saw its future as a beef animal, and it was soon put on the coastal plains of the United States and the interior Llanos of South America to be crossed with *Criollo*, Portuguese and native American stock.*

**Criollo, Native American and* Protuguese cattle are all identified as separate types by John E. Rouse in *The Criollo, Spanish Cattle in the Americas*. University of Oklahoma Press, 1977. *Criollo* are described as cattle descended from early Spanish breeds, while *Native American cattle* refer to the mixed progeny of European cattle brought first to the British colonies, and *Portuguese cattle* refer to direct progeny of cattle from Portugal, islolated in Brazil by geographical barriers.

Guzerats were exported to Brazil as early as 1870, and fairly large consignments began arriving there in 1914.* Besides being crossed with Portuguese stock, they were used for pure-breeding as well as for developing the Indu-Brazil breed of cattle.[8]

Some descendants from these same Guzerat cattle came into the United States from Brazil by way of Mexico in 1924 and 1925. (See Chapter IV, "Those Brazilian Brahmans.")

Another Indian breed used by early American cattlemen that contributed significantly to the perfection of the American Brahman was the Gir.*

The native home of Gir is the gentle Gir hills and the Teak, blackwood forests in the south of the Kathiawar peninsula, but they are indigeneous to many parts of India -- covering an area of approximately 7,000 square miles. Some are found in the relatively high elevations of the Deccan Plateau where the annual rainfall is generally less than 10 inches. And they inhabit regions where there may be a 90° F. fluctuation in temperature.[10]

Besides those cultivators with small farms, Gir cattle in India are largely raised by professional breeders who depend almost entirely on natural pastures, having their cowherders move them about the countryside like early frontier-American herders did.[11]

"The most noticeable characteristic of the Gir is a very prominent and broad forehead, which forms a heavy, boney shield covering part of the head." It has ears that are markedly long and pendulous, and the horns are peculiarly curved ending in a

*Quoting from A. A. Santiago (1978) James O. Sanders of Texas A&M says that ". . .small numbers of Zebu (*Bos indicus*) cattle were taken to Brazil during the 17th and 18th centuries and were crossed with Brazilian cattle of Iberian origin . . . From 1890 to 1921 over 5,000 Zebu (*Bos indicus*) cattle were brought into Brazil from India."

* *Gir* is sometimes spelled *Gyr*; however, this particular breed is associated with the *Gir* Forest in India which is normally spelled with an "i". Most early documents, such as Sir Arthur Oliver's *A Brief Survey of Some of the Important Breeds of Cattle in India*, 1938 and Joshi and Phillips's *Zebu Cattle of India and Pakistan*, 1953, also use the *Gir* spelling.

The Guzerat or Kankrej *Bos indicus* **breed has been highly prized in Asia for thousands of years because of its great strength and durability. It has also played an important part in the evolution of the American Brahman.**

fine taper. One of the most appealing qualities of the Gir to American breeders has been its smooth comformation. Its docile temperament has aided the U.S. cattlemen in incorporating this quality in the American Brahman. It is also believed that the large, broad forehead, the short muzzle and the straight profile of the American Brahman is at least in part the result of having used the Gir breed.[12]

Gir breeds have been particularly popular in Brazil where they have been utilized to produce beef on grassland feeding alone. In range tests made at the Experimental Breeding center of Sertavzinho, it was noted that the dressed meat amount from Girs range fed there amounted to 62%[13]

This Gir bull, named *Arauto,* was brought to the United States from Brazil in 1946. It is considered to be the progenitor of many of today's American red Brahmans.

It is in Brazil and other parts of Latin America that all *Bos indicus* breeds are referred to as *Zebus.**

While there are no large herds of purebred Gir cattle in the United States today, a few breeders through careful selection have approached the true Gir type.[14] But important to the U.S. breeders of the American Brahman, has been the role the Gir has played in producing a red Brahman strain or type. Although

*The etymology of this word is difficult to establish; however, one source traces it to a Monsieur Buffon, a French naturalist, who changed the spelling of *Zebr,* a type of Polish ox, to *Zebu* to describe the beasts of burden in India. (See Joe Dial, ABBA Papers, Houston, Texas.)

not considered a separate breed from the American Brahman, some breeders and buyers have preferred the red coloring. Part of the reasoning for this is that red Brahman bulls transmit a high degree of uniform red color when crossed with other cattle.[15]

While red colored Indian cattle were in some of the earlier American importations, U. S. breeders have developed much of their genetic pool for red Brahmans from progeny derived from Brazilian imports. However, some red Brahman breeders are insistent about head conformation, wanting to make sure their reds do not have Brazilian Gir features. One outstanding Brazilian bull that played a large role in the development of the American red Brahman was *Red Dick*, imported by way of Mexico in 1924. Another Gir bull named *Arauto*, from a 1946 importation, is considered to be an important forefather of many American red Brahmans.[16] (See Chapter VI, "Some Prefer Reds.")

The "poor man's cow", or the Nellore (also Ongole) breed of Southern India has also had an important influence on the development of the American Brahman. They are probably the most prevalent single breed of the two-hundred million cattle that inhabit India. Most are found in small village herds whose appearance there is as timeless and as classic as the dark skinned men who attend and work them.[17]

They originated in the Ongole region of the Madras State in an area not dissimilar in physical geography and climate to the Southern Coastal Plains of the United States. Competitive show classifications for this breed are probably older than any of the *Bos taurus* or other *Bos indicus* breeds, going back to the 1600's.[18]

Nellores were steadily brought into the United States beginning with the earliest importations. Early Indian cattle brought by animal shows -- many of which found their way to American breeders -- were probably of the Nellore type. And in the largest of the early U.S. importations, the Borden-O'Connor importation of 1906, six of the high quality bulls were Nellore. These six bulls came to play an important role in the course of time toward the creation of the American Brahman.[19]

Ongole or Nellore cattle are one of the most popular *Bos indicus* breeds in the world. Nellore females have contributed significantly to the development of the American Brahman.

It should be remembered that while the blood from some of these early importations may have been practically bred out, their mere presence and use served to remind cattlemen of the value and importance of preserving *Bos indicus* blooded stock.

It was also during the same year of the Borden-O'Connor importation that a large quantity of Nellores were brought to Brazil. Some descendants of these also undoubtedly reached the United States during the U.S. importations from Brazil, and thus also contributed toward creating the American Brahman.[20]

While Nellores have long been noted for their hardiness, their resistance to insects and disease and their economy of care, some American Brahman breeders have been less than enamored with their temperament and their beef producing

Krishna Valley *Bos indicus* cattle were popular in the United States because of their weight and size. Although this breed was developed fairly recently, it has also played an important role in the development of the American Brahman.

qualities. However, 52 head of pure Nellore bulls were recently imported into the U.S.A. from Brazil.[21]

A breed developed during the last part of the nineteenth century, the Krishna Valley breed, has given the American Brahman some of its most enduring and estimable qualities. It is believed that a group of rich Indian Rajahs established this breed during the last decades of the nineteenth century. By utilizing top grade Gir and Nellore cattle, with possibly some Mysore and Guzerat strains, they produced an animal not unlike the modern American Brahman although there has been a wide variation in the characteristics of this breed. The first of this breed brought to the U.S.A. presumedly came with the 1906 importation, and they were selected in India by A. P. Borden for beef characteristics. Generally they were massive and powerful in appearance and strength.

In India and Pakistan there are over thirty breeds of *Bos indicus* cattle, but the Guzerat or Kankrej, the Krishna Valley, the Nellore or Ongole and the Gir breeds have contributed the most toward the development of the American Brahman.

The *Indu Brazil* breed was developed in Brazil principally by crossing purebred Guzerat and Gir breeds. A sizeable number of *Indu Brazil* cattle were brought to the United States between 1924 and 1946.

Their name was derived from the Krishna River Valley.* It was here and in the upland watershed regions of Ghatprabha and the Malprabha Rivers that they evolved. Much of the forage here is of an extremely nutritious variety, however, rainfall is

*The Krishna River which rolls for 800 miles through India's Deccan Plateau was named for the Hindu god Krishna, creator of the universe, who interestingly grew up among cattle and cattle herders, and the stories of his childhood with them are favorite Hindu tales. See L. Schulberg, *Historic India*, Time Life, Inc., 1968.

again very seasonal and the Krishna Valley cattle in India are normally stall fed part of the year.[22] ±

While these four breeds, the Guzerat, the Gir, the Nellore and the Krishna Valley, have been the basis for the creation of the American Brahman, it should be remembered that some of the latest and largest of the U.S. *Bos indicus* importations came from Brazil.* These cattle were thus influenced in some ways by Brazilian range conditions. It is also possible that their bloodlines were effected to some slight degree by descendants of Portuguese cattle that were possibly bred with these *Bos indicus* cattle.[23]

Some of the Brazilian imports that came into the Unites States in 1946 were classified as Indu-Brazil. This breed actually took form in Brazil mainly during the 1920's and was once called the "Indu Uberaba" by Brazilian breeders because it was developed in the region around Uberaba in the state of Minas Gerais, Brazil. "The Indu-Brazil was developed from a crossbred base that involved primarily Gir and Guzerat but which also included Zebu animals of other breeds, especially Nellore."[24] These cattle have generally been big cattle with select mature bulls ranging in weight from 2,000 to 2,600 pounds. They have also been characterized by their large ears with their other features generally intermediate to those of Gir and Guzerat.[25] "In the early years of the breed, both red and grey animals were common; more recently, most animals have been grey."*

Four of the most important Brazilian bulls that were imported via Mexico in 1924, were *Aristocrata*, sire of the famous *Manso*

*"Over 80% of the Zebu cattle that have been imported into the United States have come either directy or indirectly from Brazil . . ." (See James O. Sanders, "History and Development of Zebu Cattle in the United States," *Journal of Animal Science*, Vol. 50, No. 6, 1980.)

*Although the Indu-Brazil breed was very popular in Brazil until about 1940, since then the Nellore and the Gir have increased in popularity. The Nellore bulls have become particularly desirable because of their tight sheaths, and the Nellore cows have been preferred in many cases because of their small teats. (See Jim O. Sanders "History and Development of Zebu Cattle in the United State," *Op. cit.*, p. 1193.)

and *Albacrata; Imperator,* sire of *King of Kaplan* and grandsire of *Emperor* 200; *Quinca,* sire of *Quinca the Great* and *Maroto,* sire of *Tippu the Great.* These early imports of 1924, and also a 1925 importation, were believed to be of the Guzerat breed.[26]

After 1946, imports of all cattle from Brazil to the United States were prohibited by the U.S. Government because of the danger of foot-and-mouth disease; however, with the opening of a quarantine station at Fleming Key, Florida, a number of Zebu animals have recently entered the United States.[27] (See Chapter IV, "Those Brazilian Brahmans.") It remains to be seen what effect these cattle, largely Gir, Indu-Brazil, and Nellore, will have on the American Brahman breed.

Native Breeds Crossed with *Bos indicus*

Because of their docility, hardiness and other desirable range qualities, Indian cattle were soon popular with a number of breeders in the United States, and they were consequently bred indiscriminately with existing native cattle.

In Southeastern Texas, which was to become the main cradle of the American Brahman, herds of the indomitable, tough Longhorn still held sovereign until the last quarter of the nineteenth century.* When Texas became a state in 1845, a number of nondescript native breeds had been brought in by the steady migration of new settlers. But the tall, boney and unfettered Longhorns had remained "uncontaminated." By the end of the Civil War they probably numbered eight million head. Rounded up, branded - but in no way domesticated - they were driven north to the railheads, often by the same men who would cross

*Although literally millions of these great Longhorns roamed the ranges of the Southwest in the nineteenth century, by the early twentieth century, they had all but disappeared from the Western landscape. However, the passing of favorable traits by Longhorns has given the breed a new popularity. The Texas Longhorn Breeders Association of America reports that today their combined registration is over 29,000 cattle. (See Scott Perry, "A Little Bit of Americana Comes to Hendersonville," *Southern Beef Producer.* Nashville, Dec. 1980.)

From *The Criollo: Spanish Cattle in the Americas*, by John E. Rouse. Copyright 1977 by the University of Oklahoma Press.

The nostalgic Longhorn held soverign for many years over the open ranges of the Southwest. By 1890 most Longhorn herds had disappeared, many having been crossed with Brahman cattle.

those left behind with the gentler *Bos indicus* cattle.[28]

Shorthorn (or Durham) cattle had been used extensively in the East and even some in Southern California before they were brought into Texas to be mixed with native types and later with *Bos indicus* types. But in 1871, Captain Mifflen Kenedy of South Texas, ". . . purchased a lot of 120 head of cattle, partly *Brahma** and partly Durham . . .". And in 1874 Captains Richard King and Mifflin Kenedy bought eight Durham bulls and fifteen *Brahma* bulls in New Orleans that they imported at Rockport, Texas. King later imported more Shorthorns for his Santa Gertrudis Ranch.[29]

*These so called *Brahma* cattle were undoubtedly crossed *Bos indicus* cattle, probably some of the popular Barrow Grade cattle from Louisiana.

Herefords were introduced into Texas in 1876, but they had no effect on the development of the American Brahman during the nineteenth century; however, Polled Durhams and North Devons were brought into East Texas from 1895 to 1900.[30] Some may have been crossed with *Bos indicus* graded cattle. By the turn of the century most of the Longhorn herds had gone the way of the American buffalo. Only a few isolated herds still existed.[31]

While the genetic ancestry of the American Brahman has been predominately *Bos indicus*, the importance of the native American types should not be overlooked when considering its early development. *Bos indicus* cattle, after all, were not originally used by American cattlemen to produce a new breed or even to preserve an old breed. These early cattlemen were simply interested in utilizing the matchless qualities of these cattle; so they crossbred them with whatever stock they had at the time. Subsequently, it was the desire to create an available and easily accessible source of pure *Bos indicus* blood that challenged some progressive American cattlemen to establish pure blooded Indian herds. Despite the fact that they were at this point interested in merely perpetuating the best *Bos indicus* traits, albeit they created instead a breed that excelled any breeds they had experimented with. Their final product lived longer, remained fertile and possessed a weight-for-age bonus on the range generally unsurpassed by any other breeds.[32]

The Word *Brahman*

Although the history of the American Brahman can be well documented, as with most appelations, there is no way of determing when the spelling that is now accepted was first used. By the mid 1800's, however, references to Brahman type cattle having spellings similar to *Brahman* began to appear in newspapers, legal documents, personal correspondence, etc.

A letter written in 1858, for instance, by Dr. A. C. Ambler of Florida and South Carolina refers to a bull he had recently purchased in New Delhi, India as "a *Brahmin* bull." Deed Records

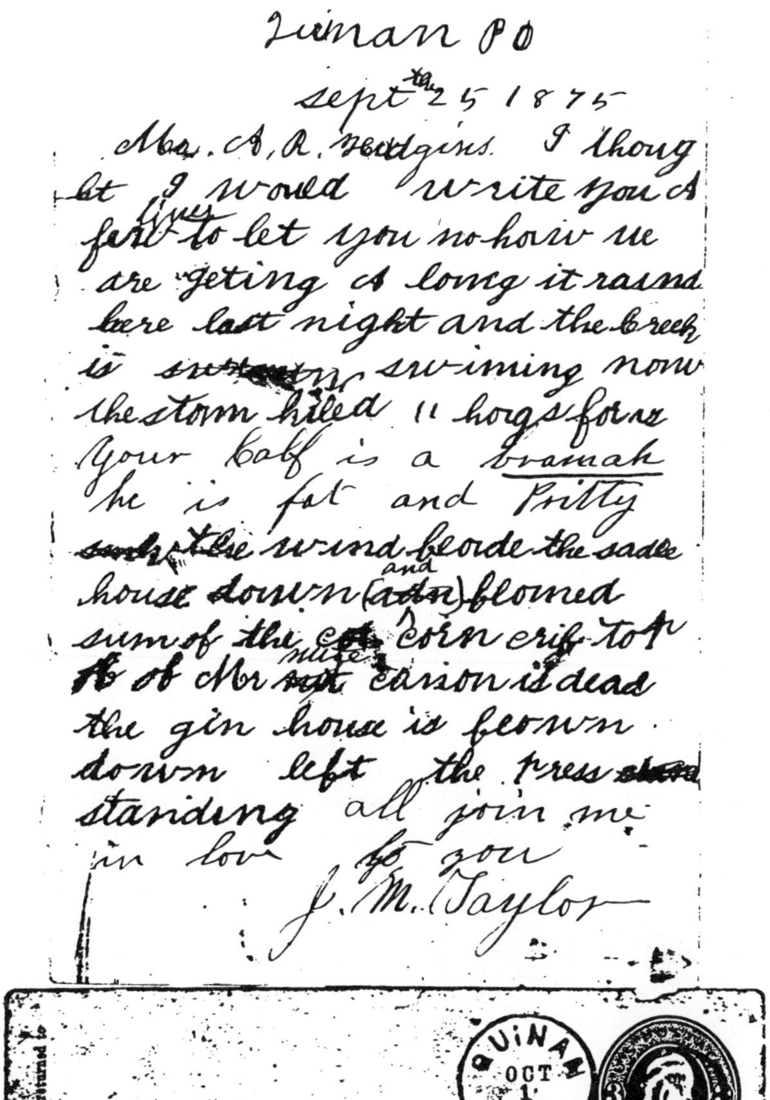

Quinan PO
sept 25th 1875
Mr. A. R. Hudgins. I thoug
ht I would write you A
few lines to let you no how we
are geting A long it raind
here last night and the Creek
is ~~swim~~ swiming now
the storm killed 11 hogs for me
Your Calf is a bramah
he is fat and Pritty
~~and~~ the wind blowde the sadle
house down ~~(an)~~ and blowed
sum of the ~~co~~ corn crib top
of Mr ~~site~~ Carson is dead
the gin house is blown
down left the press ~~stan~~
standing all join me
in love to you
J. M. Taylor

A letter addressed to Mr. A. R. Hudgins of Chappel Hill, Texas from a J. M. Taylor of Quinan, Texas postmarked September 25, 1875, contains a reference to a "bramah" calf. The calf is described as being "fat and pritty."

for Comal County, Texas in 1869 show the transfer of "a considerable number of livestock, including one *Brahma* bull" to one Charles Power, who interestingly, was a husband of Antoinette Lea, sister of San Houston's last wife. The spelling *bramah* appears in a letter received by A. R. Hudgins of Chappel Hill, Texas dated September 25, 1875. And in 1878, a Tampa, Florida newspaper account makes reference to *Brahmin* cattle that were exhibited at a state fair in Gainesville, Florida that same year. One of the most common denominations used by early cattlemen was *Bremmer.* Shanghai Pierce was known to have used this term.[33]

There was no formal adoption of the word *Brahman* until the American Brahman Breeders Association was organized in 1924. Cattlemen attending this organizational meeting wrestled with the question of what to name this American *Bos indicus* breed that J. W. Sartwelle referred to as ". . . an entirely new breed of beef cattle." Sartwelle, the first secretary and early driving force of the new association, was, indeed, the one who suggested the word *Brahman*. It was accepted by those present at that first historic meeting ". . . designating the cattle in question (and) incorporated in the name of the registry association and recorded in its charter."[34]

Sartwelle later related that as he had studied the history of India and its cattle, he had searched for an appropriate name, ". . . for the new (American) breed which we were developing." He recalled, "*Brahma* was inaccurate, and really there was no such word as *Brama* (which had also been considered.)* I didn't like the Latin suffix the British used, *Brahmin*, so I decided on the English suffix and made it *Brahman*."

Thus the word *Brahman* was the first word to be used in any formal sense in reference to North America's first distinct *Bos indicus* breed. It has stuck ever since.[35]

*Sartwelle's belief that the word *Brahma* was inaccurate presumedly was based on the fact that Hindu priests in India had used the name to refer to certain bulls they branded which became idols, allowed to roam at will. Although these cattle were also *Bos indicus* cattle, they were different from the American Brahman. (See Lloyd Clyburn, ABBA Papers, Houston, Texas, 1958.)

A common, but still beautiful sight today is a herd of purebred American Brahman cattle. This scene is somewhere on the Gulf Coastal area of the United States but could be almost anywhere in the world.

A Standard of Excellence

Shortly after the American Brahman Breeders Association was organized at Houston, Texas, its members set about to establish a Standard of Excellence and a system of registration so that the American *Bos indicus* type could be standardized and improved. And it was decided that any cattle qualifying for registration through the ABBA would henceforth be known as American Brahman.[36]

Through the years, this standard would be altered some to meet particular market demands, but it has remained basically the same. Below is the Standard of Excellence as it exists today:

STANDARD OF EXCELLENCE FOR AMERICAN BRAHMAN CATTLE

Description	Score Bull	Score Cow
I. GENERAL APPEARANCE		
A. Size and Weight-Well developed according to age. Bulls 1600 - 2200 lbs. at maturity in good flesh. Cows 1000-1400 lbs. Body frame should have ample depth, width and length regardless of condition	10	10
B. Form-Massive, long, broad, moderate depth, balanced. Straight back with a slightly rounding rump. Any appreciable dropping off from hips to region of crops or hump undesirable. Bottom line straight except for sheath in bull and navel in cow	8	8
C. Quality-hide soft and pliable, of medium thickness, hair of medium texture, soft. Bone ample in substance clean cut and strong. Muscling long, well defined	4	4
	22	22
II. BODY FORM OR CONFORMATION		
A. Shoulders and Chest-Moderately oblique, smooth, broad on top and covered by hump. Brisket not prominent. Chest wide and deep, good width on floor	8	8
B. Body		
1. Back and Ribs-Ribs well sprung from backbone, arched, with ample length to give moderate depth to body. Symmetrically joined to loin and crops. Excessive depression behind shoulders objectionable. Short middle and excessive depth of rib not desirable. Back and loin uniformly wide and gently sloping to sides (like quonset roof) when viewed from rear. Well covered with thick, natural muscling. A sharp angle of back extending above hooks objectionable	9	9
2. Loin-Broad, thick, level and firm, blending smoothly into back and rump	8	8
C. Hindquarters		
1. Rump-Long, wide at pins and slightly rounding toward tailhead. Smoothly joined to loin (steep slope serious discrimination). Tailhead smooth	8	8

2. Hooks-Slightly below level of back, medium in width, well laid in		1
3. Round-Broad, thick, full and deep, extending well down to hock	8	8
D. Feet and Legs-Moderate in length, straight and squarely placed. Bone with ample substance, strong and clean, tapering into well-formed, dense joints. Hind legs perpendicular when viewed from rear but slightly inclined forward below hocks; muscular above hocks. Strong, moderately sloping pasterns. Toes uniform, ample in size and straight, heel deep. Walk straight, strong and active	7	7
E. Muscling-Animal should show indications of superior muscling: front legs set wide, shoulder thick, forearm and stifle well muscled. Rounding over top. Rounds thick and full with widest point half way between rump and hock when viewed from rear. Animal should stand square, walk with hind legs set well apart while traveling true. Good length from hook to hock and from pin to stifle. Bone relatively heavy	7	7
	55	56

III. BREED AND SEX CHARACTERISTICS

A. Color-Grey or red of varying shades predominate. Brindle, gruella and true white (albino) are disqualifications. Muzzle, hoofs and switch black. Horns dark.		
B. Head-Sex should be expressed by the head. Bulls masculine and female feminine. Face moderate in length, muzzle full, nostrils wide and open, lips dark. Eyes mild and full with good width between them. Horns set wide at base, thick and medium in length. Horns of cows should be thinner than those of bulls . .	4	4
C. Neck and Throat-Neck moderately short, muscular in bull; neat in cow, blending smoothly into shoulders. Throat clean on sides, but with moderate development of dewlap	2	2
D. Hump-Bulls should possess hump of ample size, located directly on top of shoulders, moderate in thickness, somewhat resembling a bean in shape and extending backwards. Females should show hump of moderate development, more oval in shape and located on top of shoulders.		

E. **Sheath and Navel-**Sheath should be of medium size and closely attached, not pendulous. Excessive development of sheath or navel objectionable	3	2
F. **Tail-**Set above pins and neatly attached to body on a level with top line or slightly below. Moderately long, black switch desirable.		
G. **Sex Character-**Bulls should possess pronounced masculinity. Females should show characteristics of refinement and femininity.		
1. **Udder-**Ample in capacity, extending well forward in line with belly and well up behind; not fleshy. Teats moderate in size and squarely placed under each quarter		4
2. **Scrotum-**The scrotum should contain two testicles, well developed, of equal size. Abnormal testicles serious discrimination	4	
H. **Environmental Adaptation Traits-**Hide densely covered with hair of medium texture, oily to the touch and capable of movement along sides. Well developed dewlap with soft, pliable skin arranged in folds extending from lower jaw to chest floor. Moderate development of loose skin under belly	4	4
	17	16
V. **TEMPERAMENT**		
Alert but docile	6	6
Total	100	100

In general appearance . . . vigorous, compact and indicating high dressing percentage, masculine and possessing an abundance of quality; the walk square, the step quick and the head up . . .

Bulls should weigh 1600 - 2200 pounds at maturity in good flesh . . . Body frame should have ample depth, width and length regardless of condition . . .[37]

And for American Brahman cows the Standard specified in part that:

It should be vigorous, compact and smooth, feminine in appearance, weighing from 1,000 - 1,400 pounds. The female should show characteristics of refinement and feminity with horns thinner than those of the bull . . .[38]

In the early years of the American Brahman Breeders Association, it was determined that before cattle could be registered as pure blood Brahmans, they had to have at least 7/8ths *Bos indicus* or Indian blood. This was later changed to a minimum of 31/32nds *Bos indicus* or Indian blood.[39]

At times when American breeders of *Bos indicus* stock tried to compete against breeders of British stock, they found themselves at a definite disadvantage. This was essentially because standards requiring compactness of conformation with comparitively shorter legs had already been established. Consequently, *Bos indicus* types in America were bred toward British specifications. Thus one of the differences with the American species of *Bos indicus* as compared to other *Bos indicus* breeds, is generally that it has been bred for greater compactness and a beefier body.[40]

While breeders of all bovine breeds will differ somewhat regarding ideal conformation, successful American cattle breeders have constantly tried to produce cattle with better slaughter characteristics. Today there is a greater demand for cattle with a high percentage of lean cuts, with moderate marbling, with longer body frames and good straight backs, and with a balanced conformation and an emphasis on muscling - particularly through the hind quarters and over the top.[41] Today the American Brahman is one of the favorite show animals in the world. And its growth in popularity can be seen in the fact that it has spread to many states in the United States, and to nearly 70 other nations.[42] Cattle producers all over the world have become aware of the universality and totality of this breed.

Thus, an American dream has been fulfilled - beyond any original expectations. And while it has met many of the needs of the past, the American Brahman's greatest contributions lie ahead as it continues to be improved and as its strengths become even more universally recognized.

For Brahmans at their best or worse
are partners of our universe.
Six thousand years domesticated
they remain uncomplicated.

Elizabeth Stilwell

2
America's First Indian Cattle

The course of civilization has been marked by a succession of revolutions, each of which has produced some dramatic economic, political or social change. While many of these reshapings have been accompanied by military violence and great social upheavals, others have been less turbulent and chaotic and more constructive and significant. One such important revolution was the domestication of cattle.

Ancient Asian Ancestors

No one knows when this revolution began, or when man first began to look on cattle as more than objects of prey and worship. There is little question, however, that as early as 6,000 B.C. cattle types similar to the species known as *Bos indicus* were being herded, milked and used as draft animals along the great Steppes of Central Asia. These early cattle raisers were independent, nomadic tribesmen whose social and political status was determined by the number and quality of the livestock they owned. As Reddy wrote, ". . . to increase the number of cows, to render them fruitful in milk and prolific in calves and to have

Scenes like this have been common for thousands of years on the great sub-continent of India. Above is a herd of *Bos indicus* cattle on a desert grazing area in Central India. This style bullock cart is probably as old as the wheel itself. Adaptations of these two wheel carts were used by many early American pioneers.

them well looked after was the dream and the ideal of the ancient Aryans*. . ."[1]

In time, a great increase in the cattle population on these once rich Steppes resulted in overgrazing, and as all herders since then, these cattlemen began to seek better pasture lands. Some migrated northward and westward, but more important to this history, are those who migrated southward into present-day

*Aryan people were those of Indo-Iranian tribes that inhabited various regions running from Mesopotamia eastward to India. *Aryan* has no real validity as a racial term. It is used, however, in reference to the Indo-European family of languages.

India. They brought with them cattle which, unlike other breeds, remained domesticated and which were the ancient ancestors of the American Brahman.[2]

Because of its long face, steep horns and characteristic hump, the early history of the *Bos indicus* breeds of India interested many people. Imprints of such cattle appeared as early as 2,500 B.C. on official Indian seals. Allowing for some artistic license, the cattle imprinted on these seals show the main characteristics of the American Brahman and other *Bos indicus* breeds.[3]

As early as 1,570 B.C., representations of Indian-type cattle began to appear in Egyptian pictographs and hieroglyphics. From Egypt, there is evidence that they spread to other parts of the Mediterranean World. By the ninth century B.C., they were common to Palestine.[4]

Indian cattle may not have been the first bovine animals to pasture the Holy Land,* but when King David sang of "the cattle upon a thousand hills. . ." in Psalms, it is not unlikely that these herds were at least in part Indian cattle.

Bos indicus cattle were also driven into Continental, Sub-Saharan Africa by Hamitic tribesmen probably about 1,000 B.C. These cattle were the ancestral branch of the important Afrikander breeds.[5]

About 500 B.C. *Bos indicus* cattle began to take on a special religious significance in India. Certain outstanding bulls were blessed and branded by the high priests of the Brahma god. (See Chapter I, p. 15)

Nearly two centuries later, Alexander the Great led his crack military units across the Mid-Asian plateaus to the Indus River.

*It is difficult to say with any certainty which cattle originally inhabited the Holy Land or that land nearest to the Eastern Mediterranean. Early artifacts suggest the *Bos primigenus* type cattle, such as the aurock, were some of the first cattle in this area. (See Rice, Andrew, Warwick, Legates, *Breeding and Improvement of Farm Animals*, McGraw-Hill, 1957.)

James O. Sanders in his "History and Development of Zebu Cattle in the United States," suggests that all Zebu or *Bos indicus* cattle descended from common nonhumped cattle that were domesticated nearly 10,000 years ago. (See *Journal of Animal Science*, Vol. 50, No. 6, 1980.)

When he and his conquerors returned, they reportedly brought 1,500 head of Indian cattle, many of which were probably trailed into Central Europe.[6]

During this same century (300 B.C.), certain officials in India, designated as "superintendents of cattle," began classifying *Bos indicus* breeds and maintaining inspection and branding programs for the protection of these animals:

> When a superintendent classified cattle as calves, steers, tamable ones, draught oxen, bulls that were to be trained to yoke, bulls kept for breeding, cattle that were fit only for the supply of flesh, buffalo and draught buffaloes; heifer calves, heifers, cows in calf, milch cows, barren cattle . . . calves that are a month or two old as well as those which are still younger; and when as he ought to, he branded them all inclusive of their calves . . . along with those stray cattle which had remained unclaimed . . . he registered the branded marks, natural marks, color and the distance from one horn to another of each of the cattle . . . .[7]

This was likely the first time in recorded history that men made any systematic attempt to classify *Bos indicus* cattle by type or use.

By the time of Christ, *Bos indicus* cattle were being used principally as draft animals in the Middle East and most of Asia where they are still usually used in this capacity.[8]

From 100 A.D. to 1,800 A.D., traders of India began bringing *Bos indicus* cattle into Eastern Africa. It is believed that descendants of these imports have been kept in relatively pure state and may be seen in Kenya today.[9]

Bos Indicus in the New World

There is some disagreement among agricultural historians as to when the first *Bos indicus* cattle were actually brought to the Western Hemisphere. Although they were certainly not commercially important in North or Latin America until well into the nineteenth century, it is possible that some Indian cattle were brought into Brazil as early as the 1600's.[10]

The earliest date assigned to North American Indian cattle importations appears to be 1835 when a Dr. Campbell and a Mr. Campbell R. Bryce of Columbia, South Carolina imported two Indian bulls and four cows from Egypt.[11] It has also been asserted that an unknown quantity of *Bos indicus* cattle were brought into Brazil a year later, in 1836.[12]

South Carolina became the site of another *Bos indicus* importation in 1849 when a Dr. James Bolton Davis received a pair of Indian bulls as a gift. According to one popular account, Davis, who had served as a U.S. agricultural advisor to Turkey, received these Zebu cattle from the Sultan of that country.[13]

It is believed that Davis was able to establish a grade herd with his bulls, but it is difficult to trace their eventual fate. Between 1849 and 1854 some of the Davis crossbreeds seem to have reached Kentucky. And in 1854, a Thomas Eades of Bourbon County, Kentucky advertised the service of a Brahman bull, presumably from Davis' stock.[14] Some evidence also suggests that at least part of Davis' grade herd was sold to Richard Peters of Atlanta, Georgia in 1854 and that the offspring from these cattle were spread pretty widely over the Southeast.[15]

During the Civil War, most of the grade cattle from Davis' original Indian bulls became victims of the war.* And although such early importations may not have had an earthshaking impact upon the American cattle industry, they served to acquaint a number of pioneer American cattlemen with the breed and its great potential.

The Barrow Grade Brahmans

Without doubt, the most significant Ante-bellum Indian importation was that received by Richard Barrow of Louisiana in 1854.

*Most of the cattle of the Southeast became victims of the Civil War -destroyed or stolen by the Union Army or slaughtered by the Confederate Army. Of the actual eleven states that seceded, Texas and Florida had the largest number of catle at the end of the war as both states were out of the major theatre of military activity.

Barrow, who was a planter living near St. Francisville, Louisiana, received two or four *Bos indicus* bulls from the British government in appreciation for instructing a Crown representative in the art of sugar cane and cotton culture. These bulls were probably shipped directly from India to be used as draft animals but were bred so extensively that their offspring and descendants became know as "Barrow Grade" cattle.* "They became quite widely know throughout the Gulf Coast Region for their rapid growth and high quality meat."[16] In 1860, Brahman-crosses, probably Barrow Grade cattle, were listed for sale in the Baton Rouge, Louisiana *Industrial Sale Catalogue*. A year later it is also believed that some of the Barrow Grade cattle were shipped or trailed to Hayes County, Texas. The transfer of a Brahman bull in a bankruptcy proceeding also occurred in Comal County, Texas in 1862. This bull may have also been of Barrow Grade stock.[17]

Most of the Barrow Grade Brahmans seemed to have survived the devastation of the Civil War, and shortly after "the Surrender," a Colonel Shannon from Galveston bought a number of these cattle - shipping them to his ranch in Texas. Shannon also sent a fine Barrow Grade bull to the ranch of J.A. McFaddin near Victoria, Texas. A real skeptic about these cattle at first, McFaddin became even more disenchanted with his humped bull when fellow Texans poked fun at the bull's slick hair and long, floppy ears.[18] But McFaddin's son, Al McFaddin, persuaded him to keep the bull, and they later both became real advocates of the breed, buying all of the Indian-blooded cattle they could find. Eventually they were even able to buy the remainder of the Barrow Grade Brahmans in Louisiana and bring them to Texas. They also acquired the John David Miller herd of Victoria County, Texas which had also been developed earlier from Barrow Grade Brahmans.[19]

*According to some accounts, these Indian cattle, though originally from India, came by way of England or Australia. And some accounts also specify that Barrow received Indian or Zebu cows as well as bulls. (See James O. Sanders "History and Development of the Zebu Cattle in the United States," *Journal of Animal Science*, Vol. 50, No. 6, 1980.)

The story of how Miller developed his commercial Brahman herd is one of the most interesting in the history of the American Brahman. Miller was only 20 years old in 1859 when he pulled up stakes in Virginia and headed for what was then the real frontier. With his family, several black cowboys and 200 shorthorn cattle, Miller traveled overland eventually reaching the plantation of the aforementioned Richard Barrow at St. Francisville, Louisiana. While Miller and his entourage were camped there, it is said that one of Barrow's purebred or high-graded Brahmans serviced one of Miller's cows. And Miller also consumated a trade with Barrow, in which he exchanged a number of his shorthorns for a number of Barrow's Brahmans.

Miller's black drovers trailed his new Brahmans along the Coastal plain to Victoria County, Texas where they were turned loose with practically no management - which was often the custom in those days of open range. It was said that Miller's stock ranged over four or five counties in Southeast Texas. Some historians believe that these were the first Zebu type cattle brought to Texas.[20]

Another important South Texas breeder who acquired Barrow Grade Brahmans was Captain Mifflin Kenedy.* Kenedy received two sizeable shipments in 1871 and 1875 that were unloaded at Rockport, Texas.[21] These cattle were evidently brought to Kenedy's Laureless Ranch near Richard King's Santa Gertrudis spread.

Other breeders who also purchased Barrow Grade cattle included Solomon Barrow and A.B. Lawrence of Goose Creek, Texas and "Uncle" Brad Fairchild of Baton Rouge, Louisiana.[22] But many cattlemen along the Gulf Coast experimented with Barrow Grade Brahmans, and by 1890 they were known from

*This was the same Mifflin Kenedy who became a friend of Richard King while serving in the Seminole Wars. Kenedy later moved to Texas in 1846 and persuaded King to join him as a steamboater on the Rio Grande. Kenedy and King went into business together and became "deeply involved in border intrigues." They also became partners of the famous Santa Gertrudis Ranch in 1860 after Kenedy's ranch was destroyed by Mexican raiders. (See William Broyles, Jr., "The Last Empire," *Texas Monthly*, Oct. 1980.)

Mississippi to South Texas. The diffusion of these cattle over the flatlands of the Gulf Coastal Plain aided greatly in acquainting cattlemen with the exceptional qualities of Indian cattle.[23]

Pure or crossbred *Bos indicus* cattle were also starting to show up in other states.

The well traveled and imaginative Dr. A.C. Ambler of Florida probably brought Indian cattle into Florida as early as 1858. In that year, Ambler wrote one of Florida's state senators asking for permission to enclose a section of public land in the Kissimmee River Valley. It was here that he hoped to upgrade the Florida scrubs that ranged there in such abundance. Ambler wrote of planning to use "four graded Brahmin (sic) bulls and a very splendid Brahmin bull . . . ." that had been imported directly from New Delhi, India. He had paid $2,500 for one of these bulls in India.[24]

Twenty-one years later one of Florida's early cattle kings, Dr. H.T. Lykes,* won a cash prize for his "Brahmin (sic) cattle he showed at the Florida State Fair" - then held in Gainesville, Florida.[25] While Lykes' show cattle were referred to as *Brahmins,* more than likely, they were high-grade Brahman

*Dr. H.T. Lykes, a physician by training, turned from medicine to the "production of cattle" during the last quarter of the nineteenth century. He was soon selling thousands of Florida scrub cattle to the Spanish of Cuba for gold coin. This operation was the genesis of the large Lykes Brothers empire. (See Joe Akerman, *Florida Cowman*, Florida Cattlemen's Association, Kissimmee, Fla., 1976.)

crosses. It was common at this time to call all humped cattle Brahmas, Brahman, Brahmin, etc.

There doesn't seem to be any documentation regarding the results of Dr. Ambler's project or of what happened to Dr. Lykes' prized *Brahmin* cattle. At any rate, it would be still nearly a half century before Brahman cattle really would begin to revolutionize Florida's cattle industry.

The Indianola Importation of 1878

As Indian crossed cattle became more and more popular among Coastal cattlemen, the percent of *Bos indicus* blood was naturally declining. The need for new pure blooded stock was partly met in 1878 when four *Bos indicus* bulls and one cow were unloaded from a Dutch ship at the then thriving port of Indianola, Texas. When the large, big eared cattle clattered down the gangway onto the unsteady docks, there was no little stir among the fascinated and somewhat bemused Texans. However there were at least two Texans in the area that seemed to recognize a certain greatness in these cattle. They were Captain John N. Keeran and the legendary Shanghai Pierce, who together bought all five animals.

John N. Keeran

Keeran and Pierce had both made money on cattle when they had trailed the great Longhorn herds to the rowdy railheads of Kansas, but they were trying to upgrade their stock now so that their cattle would have a greater chance against diseases and insects and could tolerate the coastal climate better.

Pierce ended up with three of the bulls while Keeran kept the other bull and the single cow. Keeran named his new cow *Lalarooke* and his bull *Breckenridge,* which he placed on his large California Ranch in Victoria County, Texas. Keeran's Brahmans were prolific, and he gave away a number of calves including one that went to the McFaddin family and another that went to Vachel Lackey of Cuero, Texas.

Descendants from Keeran's *Lalarooke* and *Breckenridge* can be seen today on the old Keeran Ranch (now the 3X and HK Divisions), in Victoria County where cattle of Indian stock have been raised for over 100 years - and where cattle owned by Mexican Ganaderos and Gringo ranchers have grazed cattle herds for over 160 years.

Pierce began breeding his new Indian stock with his lanky Longhorns and with an existing herd of crossbred Brahmans he had picked up soon after being mustered out of the Confederate Army.* He was soon producing cattle that were very resistant to Spanish or Texas fever and other diseases that plagued American cattle. He was always a great promoter of *Bos indicus* cattle though he made his first money on Texas Longhorns. In time he also acquired high-grade Indian cattle from the firm of Frost and Montgomery.[27]

The Frost-Montgomery Brahmans

During the 1880's two livestock commission merchants, Albert Montgomery of New Orleans and J.M. Frost of Houston, became interested in the graded Indian cattle that began to appear more frequently on the New Orleans market. When a load of Brahman-blooded heifers that were shipped to Houston were rejected by some of Houston's butchers, Frost pastured them on his own ranch near Houston. Frost wired Montgomery not to send any more of the "damned slick haired cattle," but he still held onto his small herd and later marvelled at their progress. He

*Shanghai Pierce was an "up-from-broncbuster rancher," who came originally from Rhode Island, soon making his mark on the Texas range. He became a cattle king on sheer gall and native intelligence. At one time during his early career, he was running 50,000 cattle while owning only 11 acres of land. Eventually, however, he controlled more than a million acres of rangeland. (See William Forbes, *The Cowboys*, Time-Life Books, New York; 1973.)

Shanghai Pierce, one of the real legends of the wild West, was an early promoter of Indian cattle. He drove Longhorn cattle up along the Chisholm Trail during the heyday of the long cattle drives and was once reportedly thrown in the Dodge City hoosegow by none other than Wyatt Earp.

soon became convinced that they ". . . were the greatest cattle ever known for the Gulf Coast."[28]

Shortly after this, Frost and Montgomery formed a partnership to breed the Indian cattle and were soon making plans to have a purebred bull imported directly from India. In 1885, Frost and Montgomery sent one of their most dependable herdsman to Calcutta, India with instructions to purchase the best bull he could find. While there, he ended up buying a large Guzerat type bull which was later named *Khedive*. This bull eventually weighed 1800 pounds.

Although directed to buy only one bull, their herdsman also decided to buy a smaller bull for himself, a Nellore which was later named *Richard III*. Frost, however, bought this same bull from the herdsman.

When these bulls were shipped into the Port of New Orleans, they were quarantined for six months, placed in the hold of an

A number of early Texas cattle breeders who started using Brahman blooded cattle, began first with Texas Longhorns. Men such as Keeran and Pierce drove great herds of Longhorns like these up along the Chisholm Trail to the railheads at Abilene, Hays, Dodge City and other wild cow towns.

old boat and handled with little care. Reportedly ". . . their feet grew out long and curled up and they got thin . . . ."[29]

During the quarantine period, Frost fought a plethora of red tape working hard to end the quarantine as the animals had shown no signs of disease. "After much persuasion and expending a considerable amount of money . . ." the bulls were finally released. Frost immediately shipped them to Houston, Texas - little more than a village at this time. It was estimated that the total cost of buying and delivering these bulls to New Orleans was over $6,000.[30]

Production records of J.M. Frost show that the Brahman bull, *Richard III*, sired for him 75 living calves within a one year period. "Some of the calves from this bull were used in the manner of driving horses." One pair of these steers were particular favorites of Mr. Frost, and he frequently drove them from his ranch to Houston - a distance of 25 miles in a traveling time of three hours.

On one occasion while driving through Houston, Frost's pair of Brahmans were startled and chased by a wild dog. Before the

J. M. Frost, Sr. of Houston, Texas became interested in graded Indian cattle in the 1880's. He and Albert Montgomery of New Orleans, Louisiana formed a partnership in 1885 and had two *Bos indicus* bulls imported directly from India. One of these bulls, *Richard III*, sired 75 living calves in one year.

bulls and wagon had come to a halt, the otherwise docile steers had crashed through the plate glass window of the J. L. Mitchel Jewelry Co. at Main and Preston and entered the store. Frost stepped down from his wagon and led the steers safely from the premises, but he was reluctant to drive them through downtown Houston again.[31]

Over the years Frost and Montgomery used *Khedive* and *Richard III* heavily, often mating them to Barrow Grade cows, then line-breeding the daughters of one bull to another.

One of Frost's regular customers had been Al McFaddin, and once when McFaddin visited his ranch he gave him a son of *Khedive*. "Another son of *Khedive* was given to Shanghai Pierce . . . and to Tom M. O'Connor . . . ." In 1889, J.D. Hudgins of Hungerford, Texas also made a purchase from Frost.[32]

Because of financial difficulties, Montgomery actually sold his share to Frost in 1892, but he later bought fifteen more head of choice Brahman-blooded heifers.[33]

Half-Bred Brahmin Bull Calves for Sale. Sired by Khedive and Richard the Third, Imported from India in 1885.

J. M. FROST,

Stock Raiser and Live Stock Commission Merchant,

49 TRAVIS STREET.

Houston, Texas, 7/23 1894

J. M. Frost was so impressed with the Indian cattle he pastured near Houston that he used a picture of them on his letterhead. The above picture included half-bred Brahman calves sired by two bulls imported directly from Calcutta, India.

When Montgomery died in 1894, his last herd of Indian cattle was purchased by A. M. McFaddin and his partner, F. B. Wood, of Refugio, Texas. In time, the two bought up as much of the original Frost-Montgomery herd as they could find, including the two original bulls *Khedive* and *Richard III*. Both McFaddin and Wood were able to keep their herds intact for many years, maintaining one of the oldest high-grade Brahman herds in the country.[34]

The original foundation herd of the Ward Cattle Company was also based on high-grade Brahman cows that were imported to Matagorda, Texas by Frost from Louisiana.[35]

Other Brahman grade cattle brought from Louisiana continued to influence the breeding of cattle in Southeast Texas through the last years of the nineteenth century. In 1883, George W. Butler, father of Milby Butler, purchased thirty-five high-grade Brahman cows in New Orleans for his ranch in Texas. Butler herded these cattle on foot all the way from New Orleans to his homestead at League City, Texas.[36]

In 1895, Josiah Dawson Hudgins, of J.D. Hudgins and Brothers, traveled to Louisiana to purchase and bring back to Hungerford, Texas more of the Indian-type cattle. They were referred to then as the "Big Eared Louisiana Cattle." Carrying approximately 50% *Bos indicus* blood, they became part of the Hudgins' commercial herd. But they had no influence on the Hudgins' Brahman herd that was later to play such an important part in the evolution of the American Brahman.[37]

A Restless Circus Bull

Despite the growing demand for Brahman stock by coastal cattlemen, the Indian cattle were still considered exotics by many people - so rare, in fact, that they were sometimes exhibited along with regular circus animals.

The Hagenbeck Animal Show of the Hagenbeck-Wallace Circus people, usually displayed a number of pure-blooded *Bos indicus* cattle along with the rest of their menagerie. In time, a

Pictured above are four of the Hudgins Brothers about 1890. Seated (L to R) are William Hudgins and Green Hudgins. Standing (L to R) are J.D. Hudgins and Alexander Hudgins. At least one member of the Hudgins family, A. R. Hudgins, was raising Brahman-blooded cattle as early as 1875.

number of these same cattle helped expand the growing Brahman foundation herds of the coastal area.[38]

One of the most celebrated of these circus Brahmans was a bull which had been given by the Hagenbeck s to a Texas cowboy named Peter Hahn. Hahn had actually received the bull calf because he had helped deliver it when the circus was playing at Columbus, Texas in 1895. Hahn's bull was born to roam. He later broke loose and began a bovine odyssey that would make him a legend among coastal cowboys and cattlemen. In a few months he had visited herds all along the Gulf Coast, and he could be traced by the handsome offspring he sired.[39]

A Texas rancher named Louis Bunge became fascinated with the tales of this Bohemian Brahman bull. And to what must have seemed a foolish undertaking to many, Bunge set out to hunt down the best of the Hahn bull's offspring. With a nifty bit of detective work, he retraced the migrations of the bull and eventually acquired about fifty heifers sired by him. Bunge also later acquired two *Indu-Brazil* bulls, *Faciero* and *Martino* which he bred to some of his Hahn heifers. "The resultant progeny became the foundation of some of today's red Brahman herds." A part of the Bunge-Hahn herd was later acquired by the A.H. Pierce estate.[40]

Al McFaddin, Promoter of Brahmans

By the turn of the century one of the principal promoters of Brahmans was Al M. McFaddin. With his father, J.M. McFaddin, Al McFaddin had almost monopolized the market for better grade Brahman types in Texas, and as he travelled about the world, he exhorted other cattlemen to use Indian cattle with the enthusiasm of an evangelist. He, in fact, became the breed's first great promoter and later the first president of the American Brahman Breeders Association. Texas cattleman, J.W. Sartwelle said of him, "Wherever [he] was, in the lobby of the Waldorf Astoria, . . . or on a steamship in the middle of the

Al M. McFaddin was believed by some to be the first great promoter of *Bos indicus* cattle in the U.S. He later served as the American Brahman Breeders Association's first president.

Seven Seas, or anywhere in Texas, he was boosting Brahman cattle."

McFaddin, like a number of other coastal cattlemen, had long become convinced that the only way to save the cattle industry on the Coastal Plains was with a heavy transfusion of *Bos indicus* blood. With an eye always open for good Brahman stock, he purchased a bull and a cow at the 1904 World's Fair in St. Louis. The new cow died before it calved, but the bull, named *Prince*, lived to become one of the breed's outstanding sires. *Prince* was light red in color and is believed by some Brahman breeders to have been the paternal ancestor of the American red Brahman strain.* One of *Prince's* most notable descendants was *Sam Houston*, the first animal registered by the American Brahman Breeders Association in 1924.[41]

The sub-tropical areas of the South and Southwest continued to be the regions where Brahmans were the most popular. However, by the early 1900's, some cattlemen from such Midwestern states as Kansas were experimenting with them on dry ranges. One such cattleman was Ike Pryor, who was having great success with crossed-Brahman he had brought in from the Pierce Estate Ranch in Texas. Pryor was particularly pleased with their ability to gain weight rapidly on native Kansas grass.[42]

*Various claims regarding the real roots of the red American Brahman have been made and continue to be made. Although the bull *Prince* may have effected the coloration of many early Indian types in America, it appears likely that cattle from the 1924-25 and 1946 Brazilian importations played a far greater role in the development of this American Brahman strain. (See Chapter I, *The American Brahman*.)

An Answer for Spanish Fever

But, perhaps, the most invaluable quality that American cattle breeders saw in *Bos indicus* types at this time was their remarkable resistance to many diseases, particularly Texas or Spanish fever. This disease had been a problem to American cattlemen since 1814, but it was not until 1886 that Doctors Theobold Smith and F.L. Kilbourne of the United States government, had discovered that Spanish or Texas fever was really a deadly protozoa, classified as *Boophilus Bovis*, carried by ticks. As early as 1861, the interstate movement of Texas cattle had been restricted as cattlemen in the Midwest and other sections became convinced that Texas trail herds carried this disease. When intrastate quarantine lines were later established in Texas, cattlemen in the eastern and mid-eastern sections of the state really suffered. The cattle industry practically ground to a halt in this region.[43]

When the real cause was discovered, the next Herculean task was to find a method by which the ticks could be effectively controlled.

This cartoon, which appeared in 1874, lampooned early American scientists who sought to discover the cause of Spanish fever. Although some cattlemen had suggested that the disease might be carried by ticks or other insects, this early cartoonist obviously thought the idea was absurd.

Some of the first Brahman-blooded cattle brought into Texas were crossed with Longhorns. The results were often remarkable, helping cattlemen to control the "Spanish fever" and other diseases. This early Longhorn, owned by the Lykes Brothers, probably had about 1/8 Brahman blood.

It remained for a Texan, Robert J. Kleberg, to steer government scientists, then working on a solution, to the right course. In 1890, he instituted a dipping program at the Santa Gertrudis Ranch.* Although the method was originally used only to curb mange and itch, he was able to also kill ticks by adding a strong solution of carbolic acid.

At the suggestion of Dr. F.M. Francis of the Texas Agricultural and Mechanical College, a number of other dip solutions were also tried, including "dynamo" oil, to rid native and European cattle of ticks.[44]

One of the earliest techniques used to curb ticks was devised on the Shanghai Pierce Ranch, which consisted of greasing cattle with lard. "It was said on the Pierce Ranch that any bull which had no Brahman blood, or a European bull which had not been greased, would die."[45]

It was not until 1914 that the standard solution, with white arsenic as its base, became the accepted dip for protecting native cattle. But the inconvenience, expense and oft times ineffectiveness of dips, greasing, etc., caused more practical coastal cattlemen to turn to Brahman-blooded stock. Besides having a resistance to ticks, these cattle were little affected by other insect pests, such as mosquitoes and flies, and accordingly they were less susceptible to anaplasmosis and other diseases spread by blood-sucking insects and external parasites. In time, they also demonstrated considerable resistance to eye cancer and pinkeye.[46]

Shanghai Pierce seems to have been one of the first Texas cattlemen to have noticed that "Bremmers" (as he always called them), did not carry ticks as native cattle did. A few years after his death, a team of scientists from the Bureau of Animal In-

*This was the same Santa Gertrudis Ranch originally acquired by Richard King in 1853. King's first purchase was a Spanish land grant called the *Rincon de Santa Gertrudis*. He paid $300.00 for the first block consisting of 15,500 acres. King's youngest daughter, Alice, later married Robert Kleberg. The couple eventually ended up with more than 800,000 acres of Richard King's holdings after he died, "and they founded the family that sustained it." (See Broyles, "The Last Empire," *Texas Monthly*, Oct. 1980.)

Once while in New York City, the legendary A. H. "Shanghai" Pierce had a life-size statue made of himself, and he designated that it be used as a grave marker for himself in the Hawley Burying Ground at Blessing, Texas. Today this monument is a well-known landmark in southeast Texas.

dustry visited his ranch at Pierce, Texas and made the following comments about his cattle:

> One of the most interesting observations was that their progeny remained relatively free from ticks while other stock in the same pasture would be literally covered with these pests. The Brahman cattle appear likewise to be less affected by other parasites and pestiferous insects . . . .[47]

With *Bos indicus* cattle and *Bos indicus* crosses, nature had provided her own way of dipping as the oil sebum secreted by these cattle was repugnant to most insects. Shanghai had his own interesting explanation though of why ticks didn't like Brahmans. He believed that when they "sweated," the salt from their perspiration was exposed to the sun and scalded the ticks.[48]

The Borden-O'Connor Importation of 1906

As the demand for Indian-blooded stock continued to increase, the per cent of *Bos indicus* blood, as had happened before, continued to decrease. By the turn of the century, there was a critical need for the importation of new Indian stock. Some new blood had been introduced through the years, by animal shows mostly,[49] but an 1884 federal law had made it almost impossible to bring in cattle from India.* And a sizeable importation was sorely needed.

Shanghai Pierce had considered making such an importation after visiting India, but he died before he was able to implement such a project. However, the executor of his estate and his nephew, Abel P. Borden, a man of great vision, was determined to follow through with Shanghai's earlier plans.[50]

Also a New Englander by birth, Borden had moved to Texas and had "trained" under Shanghai. Like his mentor, Borden was full of enterprise and initiative. After reading a report written by David Fairchild on Australian Brahmans,* Borden decided he would make a special trip to India and bring back a sizeable herd of Indian cattle. Early in 1904, he traveled to Washington, D.C. to see Dr. A.D. Melvin, then chief of the Bureau of Animal Industry of the Department of Agriculture. He was accompanied by none other than Dr. Fairchild, who helped Borden present his case. Both men made a convincing plea, and a special permit

*The chief exception to this law was for the importation of Indian cattle for menageries and zoological gardens.

*Fairchild later confessed that he thought Borden was the only cattleman that had taken his report seriously. Fairchild's report was published by the U.S.D.A. in 1904.

Abel Pierce Borden, nephew of "Shanghai" Pierce, was a man of great vision. If it had not been for his success in bringing new Indian cattle to the United States in 1906, the development of the American Brahman would certainly have been delayed.

was granted to Borden; however, Melvin imposed a number of tight restrictions on his activities in India and his handling of the cattle.

First of all, it was required that Borden take along a U.S. Government animal inspector with him to inspect any cattle for disease that he might purchase in India. Furthermore, it was specified that before any Indian cattle could be brought into the continental United States, they would have to be quarantined for a period at Simonson's Island in New York harbor. All expenses, of course, would have to be paid by the Pierce Estate.[51]

Borden was excited about the prospect of bringing pure blooded Indian cattle to the Pierce Estate ranch, but being the good businessman that he was, and appreciating the old adage that "behind every successful man there is a smart lawyer," he decided to consult his attorneys, Proctor and Vandenberg, before pro-

Thomas M. O'Connor, a Texas cattle king, played a vital role in the 1906 Importation by going into partnership with A.P. Borden and helping him underwrite the cost of this ambitious undertaking. O'Connor received eighteen of the cattle that were eventually brought to Texas. When this picture was made, O'Connor was only 17 years old.

ceeding any further. The financial outlay, needless to say, would be considerable.*

As it turned out, Proctor and Vandenberg were able to interest Thomas M. O'Connor in the project at this point. O'Connor, a wealthy cattleman, was one of those Americans who had constantly labored to channel the stream of destiny, and he became a partner in the venture. Because of bad health at the time

*It should be mentioned that Borden's reason for going to India was twofold. First of all, he wanted to purchase a herd of top grade Indian cattle, but he was also interested in observing the culture of tea there as Shanghai Pierce and he had earlier thought of raising tea in East Texas. (See Chris Emmett, *Shanghai Pierce, A Fair Likeness*, p. 302.)

though, O'Connor decided against making the trip to India.

It was almost two years before Borden could get underway. But in the spring of 1906, he left for Bombay where he later made connections with a U.S. Government inspector, Dr. William Thompson. In many ways Dr. Thompson was ideally suited for the task. As a veterinarian and a Texan, he was interested in promoting that state's beef industry, but more importantly, he had served in the Philippines before where he had

Abel P. Borden and Dr. William Thompson travelled hundreds of miles over India in an ox-drawn cart like this one above. This picture was actually taken by Borden while he was in Bombay in 1906. Borden took great care in his selection of *Bos indicus* cattle, buying only those that had good beef qualities and were in excellent health.

gained experience with Asian cattle. There was no question in his mind about the importance of this enterprise.

Borden's and Thompson's experiences in India were like a series of Kipling tales, and have been compared to a "Frank Buck wild animal expedition."[52] Perhaps, it would have been simpler if Borden had confined his cattle purchases to the immediate Bombay area, but he was determined to make a more extensive tour of the vast subcontinent so that he could locate the best available cattle. There were also a number of regions that were suspected of having diseased cattle, and his permit prohibited him from purchasing cattle in these places - not that he would have wanted to anyway.

Leaving Bombay, Borden and Thompson traveled over the rutted back roads of India in an ox-drawn cart. When they reached Miraj, India, Borden had to bargain for two days before he was able to purchase 20 bulls from local villagers.

Buying cattle in India was a painful and oft times frustrating process. Because of their religious significance, Hindu elders were usually opposed to selling the village cattle. Ownership transfers were doubly difficult because several men would often own shares of the same animal. With some bulls, a dozen different men might own a part interest. And once the sales were finally completed, it was necessary to ship the cattle in disinfected rail cars all the way back to a farm at Poona, India. Nine cattle were eventually bought at Ahmadabad. And six others were purchased at Madras, approximately 600 miles from Bombay where they had first embarked.[53]

After many days of exhaustive travel and endless bartering, Borden had managed to purchase forty-six bulls, two cows, one heifer and two calves. Although he had hoped to buy a herd of 100 cattle, time and resources dictated that he cut his original goal short. But he had ended up with prime Indian cattle, which included twenty-two Krishna Valley bulls, nine Guzerat bulls, six Nellore bulls and at least one Gir bull.[54]

One of Borden's greatest problems in handling his newly acquired cattle was in feeding them. Again, in order to meet specifications set forth in his permit, he had to obtain hay from areas where no untested Indian cattle had ranged. He was also forced into using a cotton press to bail his hay, one that "had never seen a sprig of grass before." And before herding the cattle to rail cars for shipment to Bombay, a trail had to be cut through the forest at Poona "so that the cattle might pass over ground that had never before felt the tread of tramping herds."[55]

Fifty-one head were eventually collected at Bombay, but there Borden ran into still another difficult problem. There were no direct sailings from Bombay to New York, so he had to ship his cattle to Hamburg, Germany first. At Hamburg they were crated individually and transshipped to a freighter destined for New York City. Added to these expenses was the cost of paying several East Indian herders that he had hired on at Bombay to take care of the cattle. It was an exorbitant outlay of cash, even for the times.

After two and a half months enroute, Borden, Thompson, the Indian herders and the *Bos indicus* cattle finally reached New York City Harbor. It was June 6, 1906. All of the cattle were immediately placed in quarantine on Simonson's Island.

After five weary months of inspection and reinspection, a trace of *surra** was found in the blood of a rabbit that was being used for testing the cattle. A few of the cattle were suspected of carrying this disease and were destroyed. When Borden's Hindu herdsmen witnesses this, they tore their hair, put on sack cloth and even threatened that there would be an uprising in India if the slaughters were not stopped immediately.

Before the discovery of "surra," Borden had returned to Texas to report on his progress to Tom O'Connor, his cousin Mrs. M.P. Withers and other parties involved in the project. Needless to say, Borden was shocked when this distressing news reached him at his home as it appeared at this time that the entire herd would soon be slaughtered. But to have to come so far at such great effort and at such sacrifice - and yet to be so close to

*Surra's scientific name is *Trypanosoma evansi*. It is an infectious anemia that affects many warm blooded animals.

Pictured here is one of Guzerat bulls purchased by Borden being unloaded at Simonson's Island after two and a half months at sea. It was not until six months after this embarkation that the *surra* quarantine was lifted and the Indian cattle were shipped to Texas.

realizing his dream, Borden was not about to give up at this point. He was soon waging a war by wire - a campaign to save the Brahmans. He and his lawyers contacted everyone they thought might help, but at first, to no avail. Finally, it was suggested to him by someone, that he wire President Teddy Roosevelt. It was believed that Roosevelt would lend a sympathetic ear since he too had cattle ranched awhile in the Dakotas, and he had been in love with the West since he was a young man. No one could appreciate the problems of the cattlemen better than Roosevelt, and, of course, no one was in a better position to help Borden.

A 200-word wire was fired off to Roosevelt which ended with this slightly jocular admonition:

> It would be a calamity to the cattlemen of the United States to slaughter those cattle just because one little rabbit caught the sniffles from one of them.[56]

The lengthy wire was followed by a request for a personal interview with Roosevelt by a Mayer L. Halff, who would represent the Pierce Estate. Halff was a Texan himself, who at the time was practicing law in New York. After also pulling a few political strings, Halff wangled an audience with the President, who in turn brought Borden's and O'Connor's cause to the attention of Secretary of Agriculture Wilson. Halff persuaded Wilson to allow a cattle barn with isolated stalls to be constructed where cattle which tested negatively could be kept. It was further agreed that after more testing, any cattle that showed no signs of *surra* could then be shipped to the Pierce Estate in Texas.

After the first frost in November, some six months since the cattle were first brought to New York Harbor, the quarantine was

Three Circle, **one of the original 1906 Borden-O'Connor imported bulls, is pictured here on the Pierce Estate Ranch. Borden was able to bring 35 healthy cattle to Texas in 1906.**

lifted, and the thirty-five surviving cattle were shipped to the Pierce Estate Ranch in Texas.[57]

From the time Borden had left Texas in the spring of 1906 until the cattle were brought to the Pierce ranch in November of 1906, the project had cost Thomas M. O'Connor and the Pierce Estate well over $100,000.[58] It was an expensive undertaking, but as events would almost immediately reveal, this single investment in time and money would produce profits and benefits to the cattle industry and to the country beyond any earlier expectations - illustrating again the importance of an available and ample supply of investment capital for individuals and groups who are interested in economic and technological progress - and who have the intrepidity to risk using venture capital.

When these cattle were brought to Texas, they were divided, with sixteen bulls and a cow going to the Pierce Estate and sixteen bulls and two heifers going to O'Connor. Among the bulls apportioned to the Pierce Estate was the famous red Gir bull named *Adam*. One bull, later named *Buck*, was a prepotent sire, and it went directly to A.P. Borden.*

The news of the Borden-O'Connor importation soon became well known in Southeast Texas, and cattlemen from miles around visited the Pierce Estate Ranch in Wharton County to see the new imports. Ranchers, such as J.D. Hudgins and his son Walter Hudgins, tried to buy part of the new stock, but Borden flatly refused to let any of the Brahmans go. At the time, his principal goal was to keep the new imports on the Pierce Estate so that he could establish and maintain a large pool herd of purebred Indian cattle.[59]

But the Hudgins', whose forefathers had settled in Texas when it was still a Mexican province, were determined to buy some of the new stock - or at least their offspring. They had been using grade Brahmans since 1875 and believed that the future

*There is some variation in the accounts of the disposition of these 1906 imports. W. S. Jacobs and J. W. Sartwelle, for instance, reported that the odd bull mentioned above actually went to Al M. McFaddin. Most accounts, however, including the Shanghai Pierce biography, indicate that A.P. Borden received this bull.

was with Indian cattle. After some nine years of trying to trade with Borden, they were finally able to buy forty graded Nellore females. These were 7/8 Zebu blooded, and they contributed immensely toward their dream of establishing a purebred Brahman herd. The Nellore cows were mated to Nellore bulls until 1924 when Walter Hudgins added three Guzerat bulls from Brazil to the J.D. Hudgins Ranch herd.

The O'Connor imported bulls were more rapidly disseminated. In 1912, O'Connor sold one of his originally imported bulls to W.D. Cornelius, of Markham, Texas, who also

Both of these halter trained bulls were part of the Borden-O'Connor 1906 Importation. This picture was taken about fifteen years later at the Pierce Estate Ranch.

later became a leading Brahman breeder.* It is also believed that a T.V. Hays obtained one of the imported pure-blooded bulls from O'Connor.

When O'Connor retired from the cattle business, by previous agreement he sold all of his Brahman females -both pure and crossed - to the Pierce Estate. By 1916, he had sold all his pure-blood bulls and other commercial stock to the Ward Cattle and Pasture Company. From the Ward company, these Brahmans were distributed to several other breeders, including Fred Locke and the Sartwelle Brothers of Palacios, Texas.[60]

From their O'Connor bulls, the Sartwelle Brothers established a two-stud line breeding program, mating the offspring of one to the other. The Sartwelles had soon established one of the best Brahman herds in the Gulf Coast area and were responsible for producing the famous bull *Manso*, often referred to as "the sire of champions."[61]

Of course, in time Borden also began to sell more Brahman cattle descended from the Pierce portion of the 1906 importation. And one year's entire production of females were acquired by Joe D. Hughes of Houston, Texas. Hughes combined this stock with Brahman stock he had acquired from the Bassett Blakley strain, producing an outstanding Brahman herd.

Descendants of the 1906 Pierce bulls were also eventually acquired by Mrs. Armel Koontz Baker, granddaughter of early Brahman breeder, Captain John N. Keeran. "Mrs. Baker was the first woman to breed these cattle independently, working the range as a man, branding and gentling her cattle."[62] And she was very active in promoting and publicizing Brahman cattle throughout the United States.*

It is hard to overstate the importance of the 1906 importation. As late as 1949, it was estimated by one cattleman that between

*W.D. Cornelius also purchased from the Pierce Estate one of the first grade calves sired by an original 1906 imported bull.

*Mrs. Baker wrote an article on the American Brahman which appeared in the popular *American Magazine* in February of 1926. She also wrote a number of widely circulated articles for newspapers and trade magazines about Brahmans.

75% and 95% of all Brahman cattle in the United States carried Brahman blood attributable to the Pierce Estate Indian cattle.[63] Such estimates are, of course, speculative and impossible to verify; however, when one considers the timing of the imports, the caliber of the stock imported and the success that Borden developed with his breeding program, the estimate sounds entirely reasonable.

Importantly, Borden and O'Connor's timing was propitious. Between 1890 and 1900 the production of beef cattle in the United States was estimated to have fallen five million head annually, while the U.S. population had increased over ten

During the 1920's and 1930's, the Sartwelle Brothers of Palacios, Texas established one of the best Brahman herds in the Gulf Coast area. Pictured above are left to right: Henry F. Sartwelle, W.L.D. "Mister Dinsmore" Sartwelle, James W. Sartwelle, P. T. "Mister Pat" Sartwelle and George T. Sartwelle. Picture was taken in 1944.

Pictured above is J. W. Sartwelle and four of the original 1906 imported bulls on the Sartwelle Brothers Ranch at Palacios, Texas. These four bulls (L to R), *El Hassen, Rambler, Three Circle* and *King George* (lying down), originally belonged to Thomas M. O'Connor. They were later acquired by the Sartwelle Brothers in 1916 from the Ward Cattle Company.

million.[64] It was not that the Borden and O'Connor importation involved many cattle, but still in just a few years offspring from these imports would span a broad horizon.

While coastal cattlemen had long recognized the importance of Indian stock for combating ticks, as well as other parasites, there had never been enough Indian cattle to draw upon. Of course, the Borden and O'Connor importation did not nearly meet the demand for *Bos indicus* blooded cattle, but it helped

Mr. A. P. Borden with some of his "pets" at the Pierce Estate Ranch. The cow in the center is from the original 1906 Importation. The other three animals are offspring from cows of that importation. Borden and O'Connor's importation was a real turning point in the evolution of the American Brahman.

immeasurably. And there would be other importations later - particularly from Brazil. In the meantime, those coastal cattlemen with Brahman and Brahman-crossbred cattle would continue to do well.

Compulsory dipping was implemented by Texas in 1917, and by 1928 it was enforced finally in Florida. But dipping had many drawbacks, and as one old timer exclaimed, "It weren't the dipping that saved the cattle business hereabouts, it were that 'ole *Bremmer* cow."[65]

The Borden-O'Connor importation was not only special because of its size, but it was also unique in that an American cattleman had gone directly to India and selected cattle on the basis of American needs and American conditions. And when Borden later put them out on the Pierce Estate Ranch, they responded dramatically on the Texas grasses. Borden lost only one of the original bulls the first year, and that was from a fight with another bull (apparently he didn't like Texas). And several bulls had reached the 2,000 pound range in a few years.

Reproduction rates were also impressive. A number of the 1906 bulls were siring 100 head of calves per year on the open range. One of these bulls that was two years old when released, sired 225 calves in three years.[66]

The majority of these bulls also demonstrated considerable longevity. In fact, when the American Brahman Breeders Association was formed in 1924, many of the original 1906 imports were still alive.[67]

Reflecting on the importance of the 1906 importation, the last directly from India until 1946, cattleman James W. Sartwelle said, "This was the real beginning of the great Texas Brahman industry."[68] Indeed, time has shown that Sartwelle's observation was accurate, and perhaps, no other single event has affected the colorful history of the American Brahman more than this one.

More Circus Importations

Because of the fear of *Surra* infestation, the U.S. Department of Agriculture continued to clamp down on the further importa-

Dr. William S. Jacobs (clergyman, cattle and horse breeder), succeeded in buying up the Hagenbeck Indian cattle in the early 1900's. Jacobs was an important early promoter of the Brahman and played an active role in the American Brahman Breeders Association. He was also a charter member of the ABBA.

tion of any Zebu cattle directly from India. However, direct importations were still permitted by circuses and animal shows. During this forty-year period, there were four Indian cattle importations of this type.

The first of these was in 1910 when Carl Hagenbeck imported two Nellore cattle for his Hagenbeck Animal Show. "These animals figured in shows and zoos and multiplied to about a dozen head and were distributed to several parties in other states."[69] Some were sold to Mr. John Price at Palacios, Texas and were later sold to William Carroll of Beaumont, Texas. Carroll was later elected the first treasurer of the American Brahman Breeders Association. A pair of these Hagenbeck cattle was also sold to Dr. T.M. Neal of Wharton, Texas and later sold to the J.D. Hudgins Estate. Another pair was sold to a man in Rushville, Indiana, while still one Brahman cow was purchased by a man in Michigan.[70]

Dr. William States Jacobs, a clergyman and cattleman from

 Houston, set out to buy all of the Hagenbeck Brahmans and

their offspring, which he eventually did, placing them on his ranch at Webster, Texas.[71]

From 1910 to 1924 there were no more importations of *Bos indicus* cattle into the United States,[72] but their popularity continued to grow, especially in the Gulf Coastal states. And 1924 would be a significant year, marked by the importation of ninety *Bos indicus* bulls from Brazil via Mexico and the creation of the American Brahman Breeders Association.

3
The Launching of a Breed Association

By 1924 *Bos indicus* cattle of various breeds had been imported into the United States for almost 100 years, and a number of cattlemen had used them with notable success, particularly when they had been crossed with British breeds and other native cattle. Despite this history, no attempt thus far had been made to organize a breeders association for the purpose of establishing standards, providing a system of registration and promoting what was quickly becoming an important new breed.

The United States Department of Agriculture had virtually ignored Indian cattle with the exception of releasing one or two short publications. Most animal scientists and cattlemen in the United States had thus far looked at the humped back cattle as mere curiosities than as commercial livestock. But a few of the breed's benefactors were ready in 1924 to undertake the task of forming an association of some kind.[1]

The chief promoter of this undertaking was one of Texas' most respected cattlemen, James W. Sartwelle of Matagorda County, Texas and later of Houston, Texas. In many ways the early days of what was chartered as the American Brahman Breeders Association was a biography of J. W. Sartwelle. Indeed, ". . . his vision, faith and labor provided the drawing force needed to carry the project through . . . . "[2]

Sartwelle was a fourth generation Texan whose family had owned lands in Texas since 1832. Sartwelle's great grandfather, Judge Silas Dinsmore, though not principally a cattleman, had owned Longhorns on land that now includes four counties in Southeast Texas.[3]

A man of great versatility, vitality and creativity, Sartwelle was a Renaissance man of sorts who liked a challenge. Tempered by the hard knocks and competition of the business world, there were few obstacles in his life that he had not been able to conquer. Although he always thought of himself as a cowman, he was also an attorney, an accountant, a telegraph operator, a real estate broker and like any good cattleman, a good practical agronomist and geneticist. An avid reader, he conducted his own early research on the *Bos indicus* breed at a time when there was little to research - at least in the way of printed material.

Sartwelle visited libraries over many parts of the country, finding little on his favorite breed. But once while at a public library in Chicago, he was elated to find a small English publication entitled *Cattle of India*. Published under the auspices of Britain's department of agriculture, Sartwelle immediately contacted them, eventually acquiring a substantial library on Indian cattle. At this time, he probably knew more about the technical aspects of Indian cattle than anyone else in the United States.

But most of Sartwelle's knowledge about Indian cattle was learned first hand. Like a number of Gulf Coast cattlemen, he had watched with much interest the merits of these cattle. And in 1916, he had the good fortune of being put in charge of a herd of Indian cattle then owned by the Ward Cattle and Pasture Company. This particular herd had been acquired from the Thomas M. O'Connor Estate, and were descended from *Bos indicus* cattle imported directly from India in 1906. These same cattle, along with the existing offspring crossbreeds, were eventually sold to the Locke Land and Cattle Company at LaWard, Texas. Using his reputation as primary collateral, Sartwelle and his four brothers, (W.L.D., Mr. Dinsmore, Henry F., Paschal T. and George T.) borrowed $25,000 in 1920 and bought this herd and placed it on the Sartwelle Brothers Ranch at Palacios, Texas.

Sartwelle had an opprotunity to work and experiment with these cattle on a day to day basis, and he later recalled, "I had become convinced that they had the permanence in the United States and that they could make a constructive contribution to the cattle economy of the United States."[4]

In the course of his research, Sartwelle was surprised to find out that there was no breed association for *Bos indicus* cattle anywhere in the world - not even in India - despite the fact that these cattle, though "not in purity," were scattered over more than half the world.*

For a number of years Sartwelle studied how other cattle and horse breed associations were organized while still continuing his study of Brahmans on his family ranch at Palacios. By studying his carefully kept records, he was able to determine the proportion of *Bos indicus* blood in certain cattle. He studied three or four years before broaching the subject of a breed association to anyone else.[5]

"The Spark That Made Me Proceed . . ."

The disdain that many of Sartwelle's fellow cattlemen showed toward Indian cattle, motivated him as much as anything toward wanting to establish an organization that would promote the fine qualities of the Brahman breed. But "the spark that made me determine to proceed," he later recalled, "was the fact that although several other breeders of Indian cattle and I would prepare animals for exhibition at the Fort Worth Fat Stock Show, they would never give us any classes there. In fact, the show wouldn't permit us to exercise our cattle in the coliseum when it was nearly empty, but wasn't even wanted by the humblest

*A herd association of Zebu cattle was presumedly organized in Uberaba, Brazil in 1923 primarily for cattle sent to the U.S. by way of Mexico. (See James O. Sanders, Op. Cit., p. 1194.)

James W. Sartwelle was in many ways the "father of the ABBA." If it had not been for his persistance and encouragement, the Association would have surely died during its formative years. Here he is pictured many years lâter in Wyoming with an antelope he shot.

cowboy to exercise his horse. And we were told that if we wanted to exercise these animals to take them out behind the barn."[6]

The feeling at the time held by many cowmen was that Brahman type cattle were too dangerous and totally unpredictable in their behavior. But those people, like Sartwelle, who had worked with Indian cattle knew that they were the gentlest of all cattle when handled correctly.

At the 1923 Forth Worth Livestock Show, Brahman cattle created such a stir that the Hereford Association called a special executive meeting "to consider danger to the great Hereford breed .. . ." But they later announced, "that there was no danger whatever to the fine Hereford cattle for reason that there was simply nothing to these alleged Indian cattle of which for the most part were of various colors, even to brindles, with the hump on his [sic] withers and . . . without pride of ancestry or without the hope of posterity."[7] Even in this age of advertising and journalistic hyperbole, one would have to search hard to find a statement so inaccurate. Ironically, in only a few years a number of Texas' prominent Hereford breeders had become important advocates of Indian cattle.

"When the essence of these thoughts were printed, Sartwelle became even more aware of the need of having an association with some sort of standards, rules and regulations, with some sort of cohesive effort and cooperation among the breeders of these cattle .. . ."[8] And he was soon contacting other Brahman breeders. A. P. Borden, who had labored so hard to bring *Bos indicus* cattle to America, did not think the time was ready. But Sartwelle was encouraged strongly by Al M. McFaddin, one of the breed's earliest and most outspoken promoters. Although McFaddin was near retirement, Sartwelle persuaded him to agree to serve as first president once an association was organized.[9]

A Modest Beginning

Sartwelle had soon contacted every breeder of Brahman cattle in South Texas that, as he said, "had a meritorious

Pierce Withers, pictured here with a Brahman descended from the 1906 Importation, was one of the charter members of the ABBA. He also served as one of the Association's first vice-presidents.

herd."[10] Although many of those he approached appeared indifferent to forming an association, a number supported the idea. At the February, 1924 meeting of the Texas and Southwestern Cattle Raisers Association in Houston, a group of interested Brahman breeders gathered in the Rice Hotel to discuss the prospects of forming such an association of Brahman breeders.[11]

Present at this first meeting were Al M. McFaddin, James W. Sartwelle, William Carroll, T. C. Dunn, Jr., J. L. Reitz, C. D. Wheeler, R. H. Wood, F. W. Neuhaus, W. S. Jacobs, W. H. Crain, Estate of Charles G. Stoddard, J. Nye Ryman, Pierce Withers, C. K. McCann, Raymond Dixon, J. S. Handy, Virgil

Brahman Bulls

The use of Brahman bulls on your range herds will result in increased weight, earlier maturity, better range quality, better feeding quality and last and most important increased market price for your calves and steers.

A strain of Brahman blood in your cows will greatly increase their hardihood and range qualities.

BRAHMAN CATTLE ARE THE COMING CATTLE AND LEADERS IN THE CATTLE INDUSTRY ARE USING THEM WITH MOST SATISFACTORY RESULTS.

Our herds are headed by full blooded Brahman bulls and we have at all times a supply of young Brahman bulls and heifers for sale.

For further information and prices address—

Sartwelle Brothers

Incorporated
PALACIOS, TEXAS

This ad was distributed as a leaflet at the 1923 Fort Worth Livestock Show by the Sartwelle Brothers. The bull is believed to be *Teddy Roosevelt,* descended from the 1906 Importation. Notice the *Brahman* spelling although this was a year before the newly formed ABBA adopted that spelling.

Parr of the USDA and Dr. J. L. Lush of Texas A&M College.

The meeting was called to order and, "upon a motion duly seconded and passed, it was resolved to form an association of the breeders of Brahman cattle in North America, for the purpose of encouraging and stimulating the further use of these Brahman cattle, and to establish a system of registration for the breed, where it may be standardized and improved.[12]

A. M. McFaddin, as he had previously agreed, became the first president of the new association. Other officers elected were Abel P. Borden (not present) of Pierce, Texas as First Vice-President; James W. Sartwelle as Secretary; W. M. Carroll of Beaumont, Texas as Treasurer; Pierce Withers of Houston, Texas as Vice-President; and William Cornelius of Markham, Texas as Vice-President. A board of directors was also elected which was made up of Al M. McFaddin, A. P. Borden, W. M. Carroll, R. J. Kleberg, Jr., James W. Sartwelle, W. S. Jacobs, J. L. Reitz, T. C. Dunn, Jr., and R. H. Wood.

As was mentioned earlier, during this first meeting the word *Brahman* was adopted as the official name of the cattle that would henceforth be registered by this association. (See Chapter I, p. 15). Before this meeting was adjourned, President McFaddin appointed an "organization committee" made up of J. W. Sartwelle, W. H. Crain and T. C. Dunn, Jr. for the purpose of preparing a constitution and a set of by-laws." The meeting was then adjourned until 2:00 p.m., March 13, 1924.[13]

It would be difficult to overstate the importance of this beginning. While its start and goals were rather modest, it was a landmark event in the evolution of the Brahman breed. In fact, it structured the framework for developing what would become the first truly American breed. And as Sartwelle later said, "It (the ABBA) was the first of its kind in the world founded on scientific premises."[14]

Through Troubled Times

Although the genesis of the American Brahman Breeders Association was a modest one, with only 16 men attending that first meeting, the strong leadership was there. And it would be sorely tested during the first decades as the infant association struggled for legitimacy, acceptance and recognition. But it was not the first time in American history that a handful of determined men met to hammer out a great idea that would become a *fait accompli.*

When the second meeting convened in March, a far reaching constitution was adopted along with a set of workable by-laws. A Proposed Standard of Excellence for both Brahman bulls and cows was also adopted as well as a score card form for appraisals. Several new members attended the March meeting. They were: Richard J. Kleberg, Jr. of the King Ranch; James Callan of Menard, Texas; Charles Armstrong, John Martin of San Antonio, Louis Le Tulle of Bay City, Texas; D. W. Williams of Texas A & M College; David Stoner, P. T. Sartwelle of Palacios, Texas and of Lake Charles, Louisiana; George Barnes, and J.M. Jones of Texas A&M.[15]

An appraisal committee was appointed by President McFaddin during the March meeting to inspect the first cattle for registration by the Association. At this time, the future of the organization lay in large measure with what success this committee might have in preparing a list of applicants for the first registrations. The first ABBA Appraisal Committee was made up of J. L. Reitz of El Campo, Texas, C. D. Wheeler and Dr. W. S. Jacobs of Houston, Texas. As everyone was in the middle of spring work, the first appraisal was scheduled for the fall of 1924.[16] But before the first appraisal took place, the metal of every Brahman breeder in South Texas would be tested as two near catastrophic events occured in the Brahman belt area. Although the need for an association had long been there, perhaps, a worse time could not have been chosen to start out on such an ambitious project as 1924.

In early December of that year the dreaded foot and mouth disease was discovered on the ranch of one of the ABBA charter members, Dr. W. S. Jacobs. Immediately after this, several of the counties where Brahman herds grazed were put under quarantine by the federal government. This initial outbreak was originally blamed on the 1924 Martin-Ruffier Importation (see Chapter IV), but Dr. Jacobs and others later contended that there was no basis of fact for this charge.*[17]

Blizzard of '24 Takes Its Toll

As if this were not enough, mother nature delivered an even more brutal blow later that same month as the worst blizzard in Texas' history struck deep into the Gulf Coastal region. It was a time when the weather was not so closely monitored as it is today. Edgar Hudgins remembers going to bed the first night that the blizzard hit:

> A warm, typical late fall rain blew against the house that night. All I put over me was a light sheet, but before dawn I was freezing to death as the rain had turned to sleet, and icy winds blew in from the north. As dawn broke, you could see the ground was covered with ice. Before it was over, we had lost 300 head of cattle, which included all breeds - the Brahman herds were not hit any harder than the other herds.[18]

When the losses were eventually tolled up, it was found that some Texas ranchers lost as much as 40% of their herds. The Pierce Estate, for instance, lost 1,500 head of cattle of mixed breeds and the Sartwelle Brothers also lost about one-half of their herd of 3,000 head, bringing the family to near bankruptcy.*[19]

As South Texas ranchers began to reel back from this devastating blizzard and "collect the pieces," another outbreak of

*It was later charged that the 1924 outbreak of foot and mouth disease was caused by "an alien plant substance brought from Japan." (See Harold Preece, "The Cattle That Changed the West," *The Real West,* May, 1968.)

*The storm of 1924 caused such serious financial loses to the Sartwelles, that J.W. Sartwell felt compelled to move from Palacios, Texas to Houston (1925) ". . .whereat he entered the real estate business while attending night law school and where he established the Port City Stockyards Company in 1931 and the Houston Fat Stock Show" (now Houston Livestock Show and Rodeo Association) in 1932.

Bulls Imported in 1924-25

THESE PICTURES CAME FROM THE FILES OF MR. FRED W. NEUHAUS.

Three Circles, Imported bull owned by O'Connor and later by Ward Cattle Co. and then the Sartwelle Bros.

Normando, Sire of the famous "Jake" bull.

Componez—Imported bull owned by F. W. Neuhaus

Neuhaus cows and O'Connor bull in 1915.

The famous "Presciosa" owned by F. W. Neuhaus and bred by several present day breeders.

Gyr Bull of Morias Importation, owned by Will Melbourne.

Imported Martin and the late Walter Hudgins.

When the dreaded foot and mouth disease was detected among Brahmans in Texas during 1924, some cattle from the Martin-Ruffier Importation were suspected as carriers; however, this was never firmly established. Pictured above are some of the bulls of that importation, plus a few from the 1906 Importation.

foot and mouth disease began in Harris County on the ranch of Dr. Jacobs. This second outbreak of aftosa was more extensive than the first, and a second quarantine was imposed on most of South Texas. Conditions at this time looked so bleak for the fledgling association that President McFaddin suggested to Sartwelle that maybe the whole idea of starting an association should be scrapped. At the time, McFaddin was getting ready to retire, and he had already turned his cattle over to his heirs. McFaddin did make a request, however, from Sartwelle, asking that a young red Brahman bull named *Sam Houston* be at least registered before the association disbanded.[20] Although Sartwelle met McFaddin's request, he had no intention of letting the association go under. As he later said, ". . . the issues were too great and important . . . ."[21]

Sam Houston, **ABBA No. 1, a red bull born on the McFaddin Ranch May 1, 1923, was the first inductee of the American Brahman Breeders Association in 1924. In March of that year,** *Sam Houston* **was quietly led into the lobby of Houston's Rice Hotel where he was sold at auction for $1,000 to "Uncle" Dick Coon of Dalhart, Texas. The money was given to the Red Cross. Within 25 years over 100,000 carefully selected Brahmans had entered the ABBA Registry.**

In 1938, Hereford breeder Dick Coon of North Texas attended an ABBA meeting in Houston, Texas where he showed pictures like this one, demonstrating the Brahmans great adaptability. This photo was actually taken near Coleville, California at a more recent time.

So it was that *Sam Houston* became the first *Bos indicus* bull to be registered by the ABBA - and what a legacy he would leave. Ironically, he was not originally purchased by a Brahman breeder but rather by one of Texas' established Hereford breeders, Dick Coon. Coon shipped *Sam* to his ranch in Northwest Texas between Dalhart and the Coldwater Creek - one of the coldest spots in the Southwest. It was here that *Sam* demonstrated that great adaptibility of his breed, thriving well on

One of the ABBA registration teams during a visit to Cuba. (Left to right) S. A. Wilbeck, Celso Gonzales, J. T. White, J. E. "Peppy" de la Torriente, Fernando Galan, Sr., Henry Koontz, Dr. Bernardo Crespo, Adolfo Dangvillecourt (in back), Felipe de la Hoz, Sr., Isias M. Cartaya, Harry Gayden, Dr. Carlos Fonts, Tony Cobb, Santiago V. Perez, Dr. Gustavo Tomeu.

the cold, arid range of Texas' Panhandle. He had soon made a convert of Dick Coon,[22] and Coon later became a strong promoter of the Brahman breed, He even attended the 1938 ABBA meeting to address the membership on using Brahmans in North Texas, showing the members photographs of Brahmans at his ranch during zero weather with snow on the ground.[23]

First Important Appraisals Begin

Although another ABBA meeting was not actually held again until June of 1931, Sartwelle and other members of the original ABBA Appraisal Committee had already begun the work of appraising the first herds. Sartwelle recalled that the first herd evaluated was on the Stoddard Ranch at Buckeye, Texas. The second was on the Jake Reitz place at El Campo, Texas. At this time, Reitz probably had the best Brahman cattle herd in the United States. The appraisals soon began to evoke a new interest in the new association, and when the 1931 ABBA meeting was called, many cattle had been appraised with approximately 2,000 Indian cattle meeting the strict requirements set forth for registration. "At the time though," as Sartwelle later recalled, ". . . the blind public knew nothing of registered Brahman cattle and would pay no premium for them. These early members who registered their cattle in the face of such odds are entitled and merit the eternal thanks of the membership. They had the faith to plow right ahead .. . ."[24]

The importance of registration soon became apparent to Brahman breeders, however, and ten years after the first Brahmans had been recorded, the cummulative total registered through ABBA had jumped to nearly 30,000. Today (1981) with over 750,000 Brahmans having been recorded by the ABBA, most breeders do not consider a Brahman a purebred unless it is registered by the ABBA.

Years of Dramatic Growth and Progress

A faith in the Brahmans was also dramatically demonstrated by J. D. Hudgins and his son Walter, when they borrowed $60,000 to pay for 30 head of *Bos indicus* cattle in 1924. That was $2,000 a head when commerical cattle were only bringing about $40 a head. Dr. W. S. Jacobs, and others who bought cattle from the 1924 and 1925 importations, were also paying top dollar for Indian cattle.[25]

The 1931 meeting had also been called to elect a new ABBA president. Cattlemen all over Texas had been saddened by the recent death of A. M. McFaddin. McFaddin had been an important force in American cattle breeding for many years and had become an outspoken advocate of the *Bos indicus* cattle in particular. He had encouraged J. W. Sartwelle with his plans to begin an association, and at a time when most men were already retired, he had agreed to serve as the ABBA's first president.

The capable J. F. Hutchins of the Pierce Estate was selected to be second president. He served from 1931 to 1935. At the June, 1931 meeting, Clinton House, of Hockley, Texas was elected vice-president; Walter Hudgins of Hungerford, Texas was elected new secretary; and Cecil House of Hockley, Texas, treasurer. New members attending this third meeting included Varis House, Russel Stanger, Lawrence Ducroz, L. P. Bunge, and C. J. Zwanzig.

CERTIFICATE OF ENTRY IN THE HERD REGISTER OF

American Brahman Breeders Association

BULL

Class AA Herd No. 1 A.B.B.A. No. 283

Name Marato

Sire:

Dam:

Dropped Registered

Markings:

Steel Gray -

Breeder:

Owner: W. S. Jacobs

Entry Fee Paid. $1 00/100

Jno. W. Sartwelle SECRETARY

A M McFaddin PRESIDENT

SEE BACK HEREOF

One of the first registration forms used by the ABBA. Notice the "AA" class designation. This was for the bull *Marato,* which was the 283 bull registered. This famous animal was owned by Dr. W.S. Jacobs of Houston, Texas.

Although the association was only seven years old in 1931, it had already been recognized by the United States Government, the State of Texas and the Republic of Cuba. In the language of the modern idiom, "It was beginning to gather momentum."[26]

Part of the business of the 1931 meeting was to modify the requirements for the registration of cattle in the ABBA Herd Book - the purpose of this being to delineate even further the characteristics of the Brahman. It was decided that seven-eighths blood would be the minimum amount of breeding in cattle eligible for registration six months from this date. It was also decided that six months hence, no bulls would be eligible for registration in the ABBA unless they were by sires and out of dams already registered with ABBA. The exception, of course, being with any individuals ". . . that may have heretofore been imported or hereafter be imported with such pedigrees as are satisfactory to and approved by, this association and the offspring of such animals."[27]

In 1932 Sartwelle and other Brahman breeders would see one of their earlier dreams realized. During that year, a competitive class for ABBA registered Brahmans was first included in the Houston Fat Stock Show, which had been recently formed under the leadership of J. W. Sartwelle. Next to the ABBA itself, the show became one of the most constructive forces in the final recognition of Brahman cattle as a distinctive and important breed. And the Houston Fat Stock Show (now Houston Livestock Show and Rodeo) became the largest and most respected show of its type in the world. In recent years, Brahman cattle have consistently predominated the other breeds in numbers at this show.[28]

Another meeting of the ABBA was not called again until 1934, but, "During this period (the ABBA) was constantly gaining in favor, both as an association and for the merit of the cattle, largely through efforts of the secretary, Walter J. Hudgins and the lady who actually was doing the work of the secretary, Mrs.S. C. Border, his sister."[29]

One of the significant orders of business completed at this meeting was the decision to provide "for the appointment by the

One of the early driving forces in the ABBA and in the Brahman industry was Walter Hudgins, pictured here with his son Edgar. Walter Hudgins served as an early ABBA secretary. In 1936 Edgar Hudgins made a dramatic effort to publicize the American Brahman at the Texas Centennial Exposition in Dallas by exhibiting a herd for four consecutive months.

board of directors of a field secretary to promote the organization."[30] New members Dave Nelson and Claude Appling were elected to the board of directors.

In 1935 Fred W. Neuhaus, who owned a ranch down on the Brazos, was elected the new ABBA president. Besides having previously served on the ABBA Appraisal Committee, Neuhaus was superintendent of the Port City Stockyards Company in Houston. His adminstration from 1935 to 1940 was characterized by conscientious and effective leadership. "These became active years of growth and progress."[31] Meetings were held on a more regular basis, providing more continuity of purpose. The first Latin American rancher, the versatile Luis Manuel Guzman Willis of Panuco, Veracruz, Mexico, joined the ABBA in 1935 along with his brother. That same year Guzman Willis' bull *El Conejo* received an "AA" rating from the ABBA.* Brahman

*The "AA" and "A" classifications were originally established by the ABBA to give special recognition to imported pure *Bos indicus* cattle and their offspring.

breeders from the West Indies, and a number of states beyond the Gulf Coastal area, had also joined the Association by 1940 as cumulative registrations topped 23,000.[32]

One of the first concerted efforts to introduce the American Brahman breed to the entire United States occured in 1936 at the Texas Centennial Exposition in Dallas. Through the efforts of Edgar H. Hudgins, Brahmans were exhibited for the entire duration of the Exhibition which lasted four months. The Hudgins Brahmans were left on constant display as no other herds were made available during the Exposition. Somehow Hudgins also managed to supervise the showing of Santa Gertrudis and *Afrikander* cattle that had been brought to the Centennial Exposition at this time by the King Ranch of Kingsville, Texas.[33]

By 1937 the American Brahman had truly found a place in the sun when Louisiana State University at Baton Rouge acquired a purebred herd of Brahmans for experimentation. These cattle were purchased from the Stark Estate and were selected by Harry P. Gayden, who later served the ABBA as Executive Secretary for 26 years. Although L. S. U. had experimented with Brahman types since 1928, this was the first purebred American Brahman herd that they or any other university or college in the world had ever owned.[34] (For more information about the L. S. U. studies see Chapter VI.)

During this year (1937) the first directors from outside the state of Texas were elected to the ABBA. They were F. J. Whitehead of Baton Rouge, Louisiana and Fred Morris of Havana, Cuba. The following year, Mrs. S. C. Border was made the ABBA's first Recording Secretary - a position she admirably held until 1948. During these first years the headquarters of the ABBA was actually located at the office of the Hudgins Ranch in Hungerford, Texas, rent free.[35]

Hot Issues Test ABBA Leather

The question of when the ABBA would eventually close its books to the registry of foundation stock and issue a herd book

Louisiana State University at Baton Rouge acquired its first purebred American Brahman herd in 1937, thus giving it the distinction of having a purebred American Brahman herd longer than any other university in the U.S.A. This picture was taken in the 1950's.

had come up in part as early as 1931, but by 1938 it had become a highly controversial issue. Those who wanted to close the registry contended it was the only way that a truly separate breed could be established - one that was different in particular ways from all other *Bos indicus* breeds in India or even in Brazil. Those opposed to such a move generally took the stand that as long as cattle were pure or near pure *Bos indicus* stock - regardless of whether or not they were offspring of animals in the ABBA registry - they should be eligible for registration. Although the actual minutes of these early meetings do not reveal how stormy and tornadic they were, the young association was agonizingly tested. Still it stood.[36] But the very fact that questions regarding policy had suddenly begun to raise such passions was evidence enough of the growing importance of the Association and of the breed. It was a far cry from those early days when J. W. Sartwelle encountered so much apathy. At any rate, the question was at least decided for a time during the February, 1939 meeting when the membership present voted to close the ABBA books to the registry of anymore foundation stock.[37]

Despite the fact that some members were disappointed, or angered, the Association continued to grow, and the number of ABBA-sponsored Brahman shows increased. It was during this time that the ABBA also started supplementing premium money for Brahman shows from its own funds.[38]

Although the question of closing the ABBA herd book had produced more than its share of polemics, a related argument of removing the classifications of "AA" and "A" in Brahman cattle from the registry was even more heated. These classifications had originally been created to provide a special designation for imported *Bos indicus* cattle, but their value was questioned by some ABBA members. Under existing rules, *Bos indicus* cattle from Mexico could be assigned the "AA" class if imported into the U.S.A. But it was felt by some that they were not as good as some domestic Brahmans classified only as "A." When the final vote was taken at the February, 1941, ABBA annual meeting, President J. W. Sartwelle felt that the issue was so volatile that he actually invited the sheriff of Harris County to attend the meeting in order to prevent any violence.[39]*

Again the leather of ABBA was tested in 1946 with the creation of a new *Bos indicus* association in Texas named the Pan American Zebu Association. The PAZA, as it is now called, was not actually organized to weaken or split the ABBA. Its reason for being grew principally out of a 1946 importation of 18 Zebu bulls made at Hidalgo, Texas. These bulls, like the 1924 and

*Although the ABBA Board of Directors voted to abolish the further use of "AA" and "A" classifications as part of the regular ABBA registration designation, an amendment to the resolution specified that it was not the intent of the resolution to depreciate ". . . the value of those cattle formally designated as class "AA," double standard full blood Brahman cattle." (ABBA Minutes, March 12, 1941.)

By the late 1930's purebred Brahmans were common over much of East Texas and Louisiana. These animals were directly descended from the 1924 and 1925 Importations and were owned by Leroy Hindes.

One of the important bulls from the 1946 Importation was *Rio Negro*. Esteban (setting on bull) and Eligio Garcia of McAllen, Texas bought several of these bulls which were brought into Texas at Hidalgo, Texas. (The gentleness of the Brahmans can be seen here.) The importation of these bulls that came to Texas from Brazil by way of Mexico, created a major stir within the Association; however, they were all eventually registered by the ABBA.

1925 importations, were from Brazil but had been moved about for several months in Mexico before they were allowed into Texas. As the ABBA Herd Book for foundation stock had been closed earlier, and as these bulls were not registered in any other country, it was not possible at the time to register these Brazilian bulls in the United States. One of the purchasers of 12 of these bulls, Esteban Garcia, later recalled, ". . . I didn't want it to be a splitting factor within the cattle industry, but it so happened that we just had to do it in order to acquire the benefit of that new blood that we imported . . . . After a few difficult times with some members of ABBA, they were finally accepted into the Associa-

In 1942, Miss Margaret King (later Mrs. Margaret Watkins), was employed by the ABBA to assist the Recording Secretary, Mrs. S. C. Border. Mrs. Watkins served the ABBA extremely well and effectively for over 30 years, later becoming Recording Secretary after Mrs. Border retired.

tion."[40] (See Chapter on Brazilian Importations for further information on the 1946 importation.*)

On several occasions breeders representing the PAZA had discussed the possibility of merging with the ABBA under the conditions that all animals registered by PAZA be placed on the ABBA books without any prior inspection of the animals by the ABBA or without the assessment of any special fees. So far this has not been done.[41]

During 1941 the ABBA also hired E. C. Stone to serve as its first field man or public relations director. The first public sale of Brahmans, sponsored by ABBA, was held in February of 1942, and that same year Miss Margaret King, (later Mrs. Margaret

*Actually this registration did not occur until the ABBA rules were changed regarding imports and foundation cattle. (Edgar Hudgins, Aug. 10, 1980.)

Watkins) was hired as an assistant to help Mrs. S. C. Border. The addition of Miss King to the ABBA staff proved to be one of the most foresighted moves the Association performed during its formative years. In time, ". . . the ABBA system would reflect [her] creative imagination and patient attention to detail."[42] ABBA President Leon Locke would later refer to Mrs. Watkins [King] as ". . .one of the four or five contemporary people who has done the most to preserve and perpetuate the breed."[43]

By the end of 1942 the cummulative registrations of American Brahmans had risen to 33,023. In order to increase representation and input from the membership, the ABBA decided to increase the size of the Board of Directors in 1942 to 15 members. At this time, there were over 150 Brahman breeders in the Association. The remainder of the 'Forties would be distinguished by the election of the first woman to the board of directors, Mrs. Wilber Webb, Jr.; the ABBA sponsorship of the first public auction, the formation of the Florida Brahman Breeders Association (the first affiliate association of the ABBA); the reopening of the herd books for *Bos indicus* females; the appointment of the dedicated and gifted Harry P. Gayden as ABBA Executive Secretary; the holding of the first National Brahman Show; the celebration of the breed's one hundredth anniversary (1849 - 1949), at Charleston, South Carolina; the election of the first non-Texan as president of the ABBA, Herman Taylor of Louisiana; and the crossing of the 100,000 cummulative registration mark for American Brahmans.[44]

Before the 'Forties had ended, one of the charter members of the ABBA, Dr. W. S. Jacobs of Houston, in a dramatic move, offered for sale some of what he referred to as his 100% Brahman blooded cattle. Since 1924, Jacobs had maintained a monopoly on a herd of Guzerat females that had been classified as "AA" by the ABBA. Breeders receiving these Brahmans were W.W. Moore, Telford Jones and W.S. Jacobs, Jr.[45]

A Silver Anniversary

1949 marked the 25th anniversary of the ABBA. These first 25 years had been eventful ones. Though sometimes seemingly

"Stack of Dollars" or *Nobille* was one of Dr. W. S. Jacobs "AA" bulls. Notice the "AA" brand on the animal's hip. For over twenty years Dr. Jacobs maintained a monopoly on a herd of Guzerat females that had been classified as "AA" cows by the ABBA.

overshadowed by the great national and internationl events of this period - the Great Depression and World War II - any honest assessment of the ABBA's accomplishments during this first quarter century would have to be laudable.

Significantly, the Association had successfully inauguarated and established a new cattle breed - one destined to be one of the most important in the history of mankind - both in its pure form and so importantly as "crossbreeding's common denominator." While it was called the American Brahman, it would become as internationl as Coca Cola, and a lot more beneficial. Because of its adaptability to environmental extremes it would become exceedingly important to emerging nations after World War II. In 1958, J. W. Sartwelle had said, "The future of ABBA looms large in the importance . . . of the cattle business in the United States, but also in the world.[46]

The image that the ABBA had hoped to establish through its first years and the goals that it sought were well summarized by ABBA President Gail Whitcomb in a statement he made during the Association's Silver Anniversary:

> An animal bearing the certificate of the American Brahman Breeders Association must be an animal of good, breeding quality. One that will unquestionably contribute to the development of the Brahman breed in its purpose of increasing the production of beef. All conscientious members of the ABBA are constantly reworking their herds to see this result. A purchaser of an animal registered in the ABBA should have full confidence that he is purchasing the best in *Bos indicus* development.[47]

Certainly this image has been established and these goals accomplished.

EARLY MEMBERSHIPS ISSUED IN THE AMERICAN BRAHMAN BREEDERS ASSOCIATION

Cert. No.	Name	City	State	Date Issued
1	**A. M. McFaddin Est.**	**Victoria**	**Tx.**	**2/28/39**
2	**James W. Sartwelle**	**Houston**	**Tx.**	**2/28/39**
3	**W. M. Carroll Est.**	**Beaumont**	**Tx.**	**2/28/39**
4	**T. C. Dunn, Jr.**	**Houston**	**Tx.**	**2/28/39**
5	**J. L. Reitz Estate**	**El Campo**	**Tx.**	**2/28/39**
6	**C. D. Wheeler**	**Houston**	**Tx.**	**2/28/39**
7	**R. H. Wood**	**Refugio**	**Tx.**	**2/28/39**
8	**F. W. Neuhaus**	**Houston**	**Tx.**	**2/28/39**
9	**W. S. Jacobs**	**Houston**	**Tx.**	**2/28/39**
10	**W. H. Cain**	**Victoria**	**Tx.**	**2/28/39**
11	**Charles G. Stoddard Est.**	**Buckeye**	**Tx.**	**2/28/39**
12	**J. Nye Ryman & Bro.**	**Houston**	**Tx.**	**2/28/39**
13	**Pierce Withers**	**Pierce**	**Tx.**	**2/28/39**
14	**C. K. McCan**	**Victoria**	**Tx.**	**2/28/39**
15	**Raymond Dickson**	**Houston**	**Tx.**	**2/28/39**
16	**James Callan Estate**	**Menard**	**Tx.**	**2/28/39**
17	**Virgil V. Parr**	**Pitch Fork Ranch**	**Tx.**	**2/28/39**
18	**Abell Bros. (Thos. W.)**	**Wharton**	**Tx.**	**2/28/39**
19	**Claude Appling**	**Louise**	**Tx.**	**2/28/39**
20	**Vose Babcock**	**Ft. Myers**	**Fla.**	**2/28/39**
21	**E. L. Bass**	**Houston**	**Tx.**	**2/28/39**
22	**Mrs. Joe Bass**	**Houston**	**Tx.**	**2/28/39**
23	**S. C. Border**	**Hungerford**	**Tx.**	**2/28/39**
24	**W. A. Border**	**Hungerford**	**Tx.**	**2/28/39**
25	**L. P. Bunge**	**Eagle Lake**	**Tx.**	**2/28/39**
26	**Lester Bunge**	**Eagle Lake**	**Tx.**	**2/28/39**
27	**Buttram and Brown**	**Oklahoma City**	**Okla.**	**2/28/39**
28	**Mrs. Mary Byars**	**Columbus**	**Tx.**	**2/28/39**
29	**Dr. T. L. Davidson**	**Wharton**	**Tx.**	**2/28/39**
30	**A. G. Dingle**	**Freeport**	**Tx.**	**2/28/39**
31	**Charles A. Ducroz**	**Brazoria**	**Tx.**	**2/28/39**
32	**Lawrence Ducroz**	**Brazoria**	**Tx.**	**2/28/39**
33	**I. V. Duncan**	**Eagle Lake**	**Tx.**	**2/28/39**
34	**Alphonse Forbes**	**Greenwell Springs**	**La.**	**2/28/39**
35	**Lanier H. Forgason**	**Hungerford**	**Tx.**	**2/28/39**
36	**Fernando Galan**	**Bartle**	**Cuba**	**2/28/39**
37	**Gibbs Bros. & Co.**	**Huntsville**	**Tx.**	**2/28/39**
38	**G. A. Harrison**	**Wharton**	**Tx.**	**2/28/39**
39	**B. B. Hart**	**Houston**	**Tx.**	**2/28/39**
40	**Cecil House**	**Hockley**	**Tx.**	**2/28/39**
41	**Clinton House**	**Hockley**	**Tx.**	**2/28/39**
42	**Joe House**	**Hockley**	**Tx.**	**2/28/39**
43	**Varis House**	**Hockley**	**Tx.**	**2/28/39**
44	**Felipe de la Hoz**	**Havana**	**Cuba**	**2/28/39**
45	**D. C. Hudgins**	**Hungerford**	**Tx.**	**2/28/39**
46	**Edgar H. Hudgins**	**Hungerford**	**Tx.**	**2/28/39**
47	**Walter J. Hudgins**	**Hungerford**	**Tx.**	**2/28/39**

48	Joe D. Hughes	Houston	Tx.	2/28/39
49	J. F. Hutchins	Pierce	Tx.	2/28/39
50	M. G. Johnson	Cordele	Tx.	2/28/39
51	D. C. Johnson, Jr.	Lindsay	La.	2/28/39
52	Tilford Jones	Houston	Tx.	2/28/39
53	J. A. Lawton, Jr.	Sulphur	La.	2/28/39
54	Louisiana State University	Baton Rouge	La.	2/28/39
55	Kent Leavitt (associate member)	Millbrook	N.Y.	2/28/39
56	Lykes Bros. Inc.	Havana	Cuba	2/28/39
57	Ted Mangum	Bay City	Tx.	2/28/39
58	E. H. Marks	Barker	Tx.	2/28/39
59	A. T. Mast	Nacogdoches	Tx.	2/28/39
60	A. C. McKowen	Jackson	La.	2/28/39
61	Morris Hassler Cattle Co.	Havana	Cuba	2/28/39
62	Henry O. Partin	Kissimmee	Fla.	2/28/39
63	Oscar Lee Partin	Kissimmee	Fla.	2/28/39
64	George Peddy	Houston	Tx.	2/28/39
65	Pablo L. Perez	Havana	Cuba	2/28/39
66	Jack Roach	Houston	Tx.	2/28/39
67	H. F. Sartwelle	San Antonio	Tx.	2/28/39
68	P. T. Sartwelle	Lake Charles	La.	2/28/39
69	W.L.D. Sartwelle	Palacios	Tx.	2/28/39
70	W. E. Shaddock	Sulphur	La.	2/28/39
71	A. N. Smith	Blanks	La.	2/28/39
72	Russell Stanger	West Columbia	Tx.	2/28/39
73	H.J.L. Stark Estate	Orange	Tx.	2/28/39
74	Diego Trinidad	Havana	Cuba	2/28/39
75	Fred W. Turner	Weslaco	Tx.	2/28/39
76	Cecil E. Whaley	St. Cloud	Fla.	2/28/39
77	H. Clay Whaley	St. Cloud	Fla.	2/28/39
78	Alfred L. Warriner	Thibodaux	La.	2/28/39
79	F. J. Whitehead	Port Allen	La.	2/28/39
80	Mrs. Lizzie Wilbeck	Hungerford	Tx.	2/28/39
81	Frank J. Williams	Houston	Tx.	2/28/39
82	B. H. Willis	Beaumont	Tx.	2/28/39
83	Carlos & Emanuel Guzman Willis	Tampico	Mex.	2/28/39
84	Winfree Bros.	Houston	Tx.	2/28/39
85	Charles Zwanzig	Genoa	Tx.	2/28/39
86	Raleigh Sanborn	Bay City	Tx.	2/28/39
87	L. M. Slone	Bay City	Tx.	2/28/39
88	Milby Butler	League City	Tx.	2/28/39
89	George Gilbert	Barbers Hill	Tx.	2/28/39
90	Monticello Drug Co.	Jacksonville	Fla.	2/28/39
91	Mrs. Wilber Webb, Jr.	El Campo	Tx.	2/28/39
92	E. C. Goldman	Victoria	Tx.	2/28/39
93	M. I. Janes, Jr.	Beaumont	Tx.	2/28/39
94	J. A. Wetzel	Rio Grande City	Tx.	2/28/39
95	B. A. Skipper	Encino	Tx.	2/28/39
96	T. I. Hester	Donna	Tx.	2/28/39
97	W. R. Lilly	Devine	Tx.	2/28/39
98	Jack Echols	Liberty	Tx.	2/28/39
99	W. A Paddock	Houston	Tx.	2/28/39

100	Dr. C. C. Conley	Raymondville	Tx.	2/28/39
101	C. R. Hudgins	Hungerford	Tx.	2/28/39
102	Edward L. Partin	Kissimmee	Fla.	2/28/39
103	Henry H. Partin	Kissimmee	Fla.	2/28/39
104	Mrs. Ethel Wilson	Brazoria	Tx.	2/28/39
105	C. K. Boyt	Devers	Tx.	2/29/40
106	E. W. Boyt	Devers	Tx.	2/29/40
107	Dr. James House Bute	Houston	Tx.	2/29/40
108	Horace Cook	Crosby	Tx.	2/29/40
109	R. S. Coon	Houston	Tx.	2/29/40
110	F. Heyd, Jr.	Iowa	La.	2/29/40
111	Homer Lord	Crosby	Tx.	2/29/40
112	R. M. Middleton	Wallisville	Tx.	2/29/40
113	A. T. McDannald	Houston	Tx.	2/29/40
114	Fred C. Schluens	Cat Spring	Tx.	2/29/40
115	E. H. Seidel	Edna	Tx.	2/29/40
116	Vernon Stegall	Franklin	Tx.	2/29/40
117	A. M. Wilburn	Cedar Bayou	Tx.	2/29/40
118	H. G. Frost	Houston	Tx.	3/4/40
119	Homer T. Johnston	Corpus Christi	Tx.	3/1/40
120	Mrs. Jimmie Haggard	Clemville	Tx.	3/11/40
121	J. O. Chapman	Chapman Ranch	Tx.	3/12/40
122	Aubrey J. Marceaux	Kaplan	La.	3/10/40
123	D. Guerra & Sons	McAllen	Tx.	3/21/40
124	W. J. Gayle	Lake Charles	La.	5/21/40
125	Fausto Yturria	Brownsville	Tx.	9/3/40
126	J. H. Tigner	Angleton	Tx.	10/7/40
127	J. V. Gates	Poteet	Tx.	10/7/40
128	J. B. Casey	Galveston	Tx.	10/7/40
129	Arthur Gilmore	Nome	Tx.	10/7/40
130	P. Woods	Mt. Belvieu	Tx.	10/7/40
131	S. A. Wilbeck	Hungerford	Tx.	10/7/40
132	Dr. James Lawrence Ducroz	Brazoria	Tx.	10/7/40
133	Rich Bros.	Liberty	Tx.	10/7/40
134	J. C. Judge	Mineola	Tx.	10/7/40
135	R. Newell Waters	Weslaco	Tx.	10/7/40
136	W. D. Partlow	Liberty	Tx.	12/31/38
137	Travis Marks	Barker	Tx.	3/1/39
138	H. H. Heidle	Kaufman	Tx.	10/21/40
139	D.J. & J.H. Phares	Woodland	La.	10/28/40
140	Mrs. J. H. Phares	Woodland	La.	10/28/40
141	Henry C. Koontz	Inez	Tx.	11/12/40
142	John Heyd	Lake Charles	La.	11/18/40
143	A. E. Shirley	Port Arthur	Tx.	1/9/41
144	Stewart Savage & Sons	Bay City	Tx.	1/13/41
145	Frank Y. Dew	DeWalt	Tx.	1/20/41
146	Glen T. Bundick	Barker	Tx.	2/1/41
147	J-M Brahman Ranch	Houston	Tx.	2/13/41
148	Mrs. Pearl Mangum	Bay City	Tx.	2/17/41
149	Wilbourn S. Gibbs (Transfer from No. 37 Gibbs Bros.)	Huntsville	Tx.	3/25/42

150	**Mack Birdwell**	**Beaumont**	**Tx.**	**3/25/42***
1277	**W. D. Cornelius** (In 1950 it was discovered that a membership had never been issued for Mr. Cornelius in the very beginning so the membership was adjusted at this time.)	**Markham**	**Tx.**	

*While this was the first membership list actually *issued* by the ABBA, this is not to suggest that the ABBA was inactive before 1939. Indeed, all the names of the founders also appear on this list, and as early as 1926-27 a number of *Bos indicus* cattle had already been registered by the ABBA. As was noted earlier, by 1931 approximately 2,000 cattle had been registered. Some members belonging to the ABBA before 1939 dropped out or died before this list was released; consequently, their names may not appear here.

The Brahman breed passed a historical milestone in April of 1973 as the ABBA registered its 500,000 animal, a bull produced by breeder Bob McAdams (center) of Huntsville, Texas. ABBA Recording Secretary at the time, Margaret Watkins and ABBA Executive Secretary Harry Gayden congratulate McAdams.

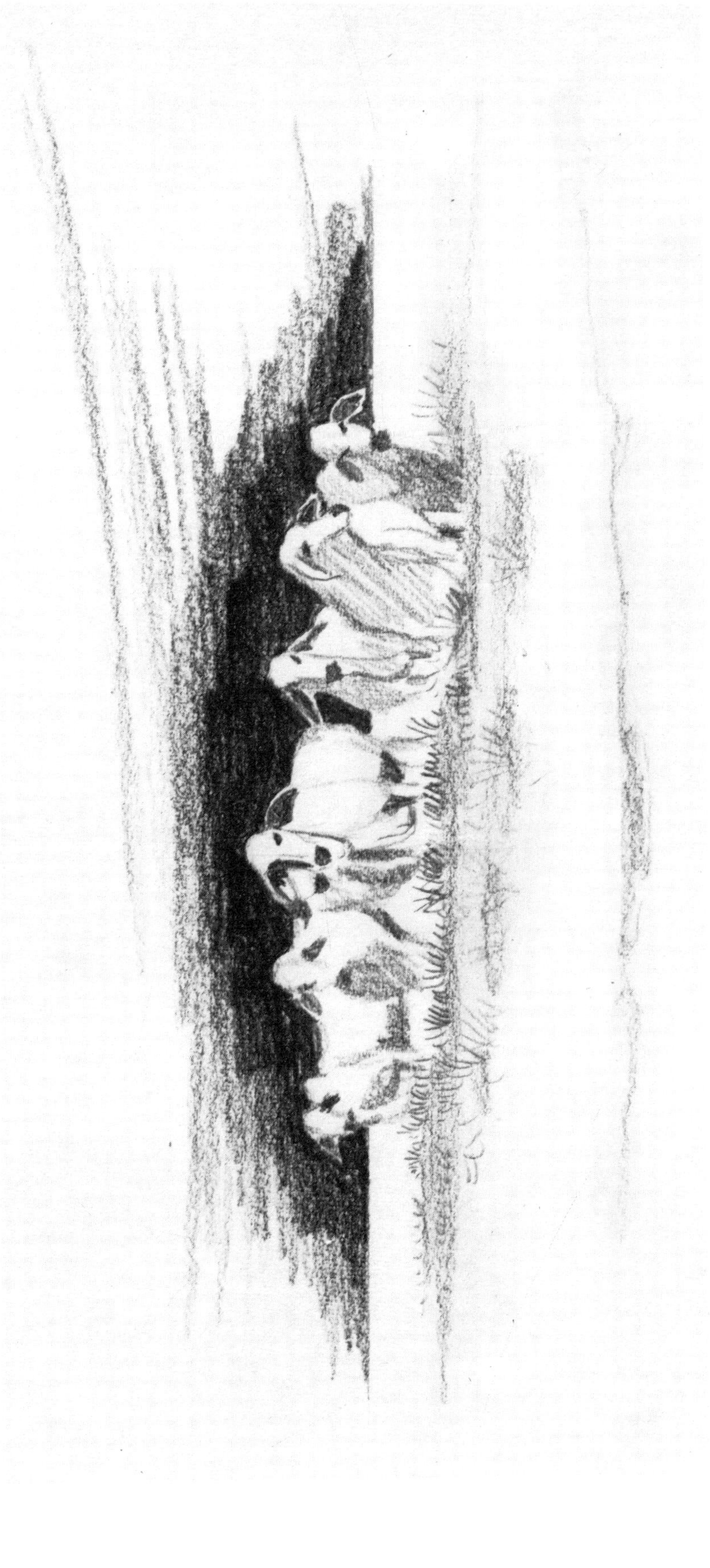

4
Those Brazilian Brahmans

The Ruffier Importation of 1924

While 1924 marked the beginning of the American Brahman Breeders Association, it was a summital year for more than one reason to the American cattle industry. Indeed, it was the year of the Ruffier Importation (sometimes referred to as the Ruffier-Martin Importation) of 90 *Bos indicus* bulls. Bred and raised in Brazil, these animals actually came into the United States through Mexico. The predominant breed of this group was the Guzerat. And there were several animals of Krishna Valley breeding as well as a few Guzerat-Gir crosses. In general the importation was made up of top quality cattle. Certainly "in terms of the most progeny as evidenced by the pedigrees of today's Brahmans, the bulls of [this] importation rank above all others."[1] No single event thus far has affected the history of the American Brahman more.

The principal author and engineer of this Herculean enterprise was Dr. Ferdinand Ruffier of Brazil. Ruffier was the organizer and manager of the Zebu Herd Book Association of Brazil,* and in 1923 he was in Mexico as an agent of the Brazilian Government to interest North American cattlemen in purchasing *Bos indicus* bulls from Brazil. Breeders in Brazil, who had been actively engaged in raising Zebus since 1870, had long known about the potential and growing popularity of Indian cattle in the Coastal Regions of the United States and Mexico.

Ruffier met cattle breeder J. W. Sartwelle in 1923 at a Mexican livestock show and expressed his desire at that time to bring a number of Zebu bulls into Texas. Although Sartwelle was naturally interested, Ruffier was warned that the 1910 quarantine on Indian cattle was still in effect. This did not, however, thwart Ruffier's plans. He was as zealous about promoting Indian cattle as were such Texas breeders as J. W. Sartwelle and A. M. McFaddin, and in 1924, ninety head of Brahmans were on the high seas destined for Vera Cruz, Mexico. But while enroute, the ship carrying the animals was diverted to New Orleans to unload a cargo of coffee and to pick up a load of flour. And as was expected, as soon as the vessel dropped anchor at New Orleans, it was boarded by agents of the U.S.D.A., who immediately impounded it for breaching the U.S. laws prohibiting the importation of *Bos indicus* cattle from Brazil.[2]

While anchored at New Orleans, Ruffier contacted A.P. Borden, A.M. McFaddin and J.W. Sartwelle, inviting them to inspect his fine cattle. Borden showed no interest in buying any of the bulls, but Sartwelle selected five, which he said he would purchase if permission could be granted to bring them into the U.S.A. McFaddin expressed interest in some of the red bulls, but told Ruffier that he didn't think they could ever be brought into the United States legally.[3]

When the destination of the bulls was finally established as Tampico, Mexico, the ship was allowed to sail. After these

*This "herd book association" was probably the *Associacao de Herd Book* or the *Sociedade do Herd Book Zebu* which is mentioned in A. A. Santiago's *O Gado Nelore, Instituto de Zootechnia,* Sao Paulo, Brazil, 1972. This association was actually not organized until 1923, specifically for registering *Bos indicus* cattle to be sent to Mexico.

When Mexican rebels attempted to seize Brahmans from the 1924 Ruffier Importation, Senor Guzman Willis of Tampico and his vaqueros were able to hide the animals until they could be driven to safety.

animals were unloaded at Tampico, Mexican rebels attempted to seize them for their army. But through the efforts of Senor Guzman Willis of Tampico, the cattle were hidden and saved from being slaughtered. Later Sartwelle visited Ruffier in Mexico at a ranch some 80 miles southwest of Laredo - insisting still that the bulls had to be brought into the U.S.A. legally. Sartwelle tentatively agreed to purchase four of five bulls he had previously selected for his friend Walter Weathersby of Del Rio, Texas. One of these bulls was the famous *Imperator*, ABBA 120, which significantly Sartwelle later bought back from Weathersby. When

One of the 1924 Ruffier imports was this Guzerat bull named *Victor.* **Walter Hudgins of Hungerford, Texas, bought this bull in Mexico and eventually placed him in a cowherd with some descendants of the 1906 Borden-O'Connor Importation.**

the Brazilian bulls were eventually brought into Texas, partly through the efforts of John T. Martin, Sartwelle bought several other bulls from this importation, including the great *Aristocrata,* ABBA 134.* [4]

John T. Martin, who owned a ranch near Artesia Wells, Texas, was particularly helpful during this time in advertising the new importations and underwriting the expenses of handling them.

Because of the strict quarantine, it was a year before the U.S.D.A. would allow any of the first bulls from the Ruffier Im-

*Other imported bulls included in Sartwelle's purchase were *Fidalgo, Castanhold,* ABBA 393, *Radio,* ABBA 392, *Tanico,* ABBA 394 and *Benficto,* ABBA 397.

portation to enter the U.S.A. Eventually they were brought up to Piedras Negras, just across the Rio Grande from Eagle Pass, Texas. The bulls were then brought into Texas in groups of 30, but even before the first group was led across by a U.S.D.A. inspector, Walter Hudgins and J.W. Sartwelle crossed into Mexico to inspect them.[5]

Hudgins bought the bulls *Victor*, ABBA 241, *Rodrigo*, ABBA 240, and *Vulgar*, ABBA 230, which were subsequently put in cowherds consisting of descendants from the 1906 Borden-O'Connor Importation. This breeding produced a Nellore and Guzerat cross which, according to Edgar Hudgins, "had the effect of increasing the skeletal size and improving the temperament of the Hudgins' Brahman herd. The calves were completely different . . . a great improvement over their dams."[6]

This same year the Hudgins started to separate their cow herds into single sire herds. Edgar Hudgins said, "[We] were beginning the move towards the purebred history, and by 1929 began to enroll cattle in the fledgling American Brahman Breeders Association."[7]

Although Walter Hudgins originally purchased only three bulls from the Ruffier Importation, his ranch eventually ended up with 30 of the 90 imported bulls; and none of these bulls were related. Some of the most important of these bulls (besides those already mentioned), which are liberally sprinkled through present-day pedigrees were *Amorim*, ABBA 44; *Martin*, ABBA 231; *Monarca*, ABBA 243; *Supimpa*, ABBA 232; *Jeremias*, ABBA 234; *Quo Vadis*, ABBA 235; *Himalaya*, ABBA 238; and *Hopkins*, no ABBA number.

Amorim was originally bought by Jake Reitz of El Campo, Texas, and then sold to the J.D. Hudgins Ranch. Described as "A very beautiful silver-grey bull," *Amorim* unfortunately broke his back during a savage storm in 1932 when a barn collapsed on him, and he had to be destroyed.[8]

Martin, ". . . a very handsome bull of Krishna Valley breeding . . .,"[9] was picked by Walter Hudgins as his choice among all the bulls he had a chance to buy in this importation.

Martin, pictured above with Mrs. Walter Hudgins, was a bull of Krishna Valley breeding which was in the Ruffier Importation.

Another of the Ruffier imports was the bull *Rodrigo* pictured here with Mrs. Walter Hudgins at Hungerford, Texas. Standing behind the animal is Walter Hudgins' brother-in-law, J. B. Forgason.

Dr. W.S. Jacobs later said that "he was so impressed with this bull that he himself leased *Martin* for six months to use on his own ranch at Webster, Texas." A very important son of *Martin*, out of *Blue Bonnet*, ABBA 1419 (daughter of *Maroto*) was *Blutin*, ABBA 2572. *Blutin* played a significant role in the development of the Sartwelle Brothers' American Brahman herd.[10] Hudgins' next choice was *Supimpa*, ABBA 232, who arrived in Texas with a bobbed tail, but was nonetheless a "handsome dark bull, of the Guzerat type."[11]

Monarca, ABBA 243, was one of the best formed bulls of the Ruffier Importation. He was half Gir and half Guzerat and nicknamed "Spotback" as he had a large Gir spot on his loin. Later he was used with the Charles Zwanzig herd at Genoa, Texas, "and he had a great deal to do with the good qualities of the Zwanzigs' cattle." *Monarca* was the half-brother of *Maroto*, ABBA 283, having been sired by *Canario*, "a distinguished and far-famed Brazilian bull in Brazil." *Maroto* had been purchased by Dr. Jacobs who had paid John Martin $500 above the going price of $2,000 for an option on him. The Pierce Ranch later had thirty of its best cows bred to *Maroto*. *Maroto*, along with *Tippu the Great*, figured in the lineage of the Brahmans presently found on the 3X Ranch of Joe and Emily Dial at Placedo, Texas.[12] *Maroto* also sired *Silver King*, ABBA 291, who was used by the Sartwelle Brothers extensively. And still another Sartwelle bull sired by *Maroto* was *Admiral* (Little Dutch) ABBA 300.[13]

Jeremias, ABBA 234, was considered by Dr. Jacobs as the "second best bull in all of the 1924 Importation, and he was of exquisite conformation, a replica of *Maroto*, only smaller." Although he was not as large as *Maroto*, *Jeremias'* progeny turned out to be extra large. He was owned by J.D. Hudgins.[14]

Another bull purchased by Hudgins from this importation included the aforementioned *Quo Vadis*, ABBA 235, a large red bull. The Hudgins' Estate did not wish to continue his line though and sold him soon after. But he later became an important sire in the development of the red Brahman. *Himalaya*, AB-

Jeremias, **pictured here with J. B. Forgason, was considered by Dr. W. S. Jacobs as the second best bull in the Ruffier Importation.**

BA 238, a big, dark Guzerat, was liked by Hudgins, and he was kept on as a herd sire.[15]

Hudgins' bull *Hopkins,* no ABBA number, was actually first purchased by Ed C. Lasater of Falfurrias, Texas, but Lasater thought the bull was sterile and returned him to John Martin. A very heavy boned, deep bodied, straight lined bull that was predominately Guzerat, *Hopkins* was one of the 30 bulls of the 1924 Importation that was eventually acquired by the Hudgins Estate. Jacobs later bought *Hopkins* from Hudgins, and the not so sterile bull sired four sons and one daughter for Jacobs. These four bulls were later sold to Joe D. Hughes of Houston, and one of them, *Hopkins, Jr.,* was repurchased by Dr. Jacobs. Jacobs in turn leased him to the Hudgins Ranch for one or two years where he was used as a herd sire. Fred Neuhaus, president of the ABBA from 1935 to 1940, eventually bought *Hopkins, Jr.* who became the sire of many excellent animals on the Neuhaus Ranch. One of these later distinguished himself in Florida.[16]

Perhaps the most travelled of these imported bulls was the celebrated *Quinca*, ABBA 49. He was first purchased from Martin by Jess Stewart and William Carroll of Beaumont, Texas, who owned him jointly. Later he was sold by Stewart and Carroll, with the rest of their registered herd of Brahmans, to C.D. Wheeler of Houston. Joe Hughes was the next to purchase *Quinca* along with Wheeler's original registered herd. From the Hughes Ranch, the itinerant bull was next sold to the Stark Estate, which was then under the management of ex-Texas Ranger and rodeo rider Dave Nelson. Dr. Jacobs next leased *Quinca* for one year during which time the bull sired *Quinca the Great*, ABBA 502. Among the famous progeny that *Quinca* produced in South Texas were *Optimus*, AA/95; *Perfecto*, AA/23, and *Quinca's Return*, A/666.[17]

Other prominent bulls purchased by Dr. W.S. Jacobs from this importation included *Preciosa*, ABBA 285; *Nick*, ABBA 4136, and *Sombrero*, ABBA 6225. Joe Hughes owned *Preciosa* for awhile but later sold him to Fred Neuhaus. While on the Neuhaus ranch, the great sire was hit and killed by a train. *Preciosa* was a large bull, and the impact was so great that the train was derailed.[18]

Nick was a darkish brown bull that Jacobs first sold to Ballich of Galveston, Texas, but Jacobs admired *Nick*'s progeny so much that he bought him back. *Nick* was later leased to Milby Butler for one year.[19]

Sombrero was almost solid red and of a fine conformation with "very wide spreading horns." Later he contributed considerably to the development of the red Brahman.[20]

Manso the Marvelous Bull

The Sartwelle Brothers' *Aristocrata*, ABBA 134, a pure Guzerat, was primarily famous for siring the prolific and renown-

Pictured here are yearling bulls from the Ruffier Importation in the sale yards at San Antonio, Texas. The young bull in the center is believed to be *Aristocrata,* **PH, ABBA 134, later recorded as the sire of** *Manso.*

ed *Manso. Manso*'s dam was the Sartwelle Brothers' cow *Penelope*, PH 109, ABBA 567, born in 1921 and sired by the bull *President Wilson* or *Wilson*, a son of the 1906 imported *Three Circle Bull.* In time, *Manso* developed an extremely impressive legacy, becoming one of the most important sires in the history of the American cattle industry.[21]

A father and son team, Walter and Edgar Hudgins of Hungerford, Texas, put *Manso* in the record books. Walter Hudgins, who had been managing the Hudgins' Ranch Brahman herd since 1915, traded four bulls to a Mrs. B.B. Gayle, a commercial breeder of Francitas, Texas, for *Manso* in 1931. Up to that time, *Manso* had not distinguished himself in any real way. Mrs. Gayle had acquired *Manso* when he was only a yearling bull from the

Sartwelles having paid $75 for him. *Manso* weighed approximately 2,000 pounds when Hudgins brought him to the family ranch at Hungerford. He was extremely wide and deep with moderate bone, good feet and legs and had what would in time become a classic head.[22] Importantly, he was range raised. His condition and conformation were not the result of any kind of supplemental feeding.

The great bull lived to be 17 years of age, producing 171 daughters and 145 sons during the ten-year period he spent at Hungerford. He sired many outstanding bulls including *Aristocrat Manso*, ABBA 1175 and *Manso, Jr. Aristocrat Manso* became Grand Champion Bull of the leading Brahman shows in Texas from 1936 through 1938. Another of his sons, *Manimoso*, was the champion of the 1946 National Show.[23]

As recently as May of 1980, the continuing significance of the *Manso* bloodline could be seen at Mexico's first National All Brahman Show held at the historical city of Ciudad de Valles. Edgar Hudgins, of J.D. Hudgins, Inc., who was invited to judge this show, reported that all of the 325 Brahmans that were exhibited contained *Manso* blood. But again, one will find many *Manso* blood-lined animals in any recognized Brahman show today.[24]

A Proven Breeding Program

The breeding program that was begun on the J.D. Hudgins Ranch was developed and dictated by a set of circumstances that sharply demonstrated the truth of how necessity is still the mother of invention. Bad farming years, killer hurricanes, rare ice storms and rock bottom cattle prices had brought the wolf to the door of a good number of coastal cattlemen. Creditors were as thick as fleas on a Mississippi hound, and many ranchers were as poor as skim milk. The only way the Hudgins Ranch could keep from going under was to come up with a superior breeding program. And with a little luck - but mostly with a lot of sweat and painful planning - the challenge was met.[25]

Manso **was 11 years old when this picture was taken. By this time he had already sired many famous progeny. He lived to be 17 years old, producing 171 daughters and 145 sons.**

Using a technique of line breeding, with *Manso* as the eventual prototype, Walter and Edgar Hudgins began to develop a long-range breeding program.[26]

Although they were committed to concentrating on a chosen bloodline, their plan was to get further away in actual kinship in each successive generation. As Edgar Hudgins later explained, ". . . to begin with you could breed a son carrying 50 percent *Manso* blood to a cow carrying 25 percent *Manso* blood, which in kinship is a half-uncle to a half-niece. The next generation you would breed another 50 percent half-great uncle to the resulting animal from a previous mating, namely, a half-great-niece that carried 37½ percent *Manso* blood."[27]

As this program was implemented, each generation was further removed in kinship but more concentrated in a blood line. Following this process and always culling unsparingly, the

One of the many outstanding products of the Hudgins breeding program was *Queen Manso Rexrdo,* **ABBA 43360, which was the 1948 National Champion Brahman Female.**

Hudgins were able to predict what the offspring would be and subsequently produced a group of sound healthy cattle. "You can go with this system of breeding forever," as Hudgins explained, "if you have adequate numbers, several hundred head are needed."[28]

According to Richard L. Forgason, D.V.M., part owner of the Hudgins cattle operation and a former ABBA president, "We are of the opinion that a good thick, deep quarter that is full in the twist is preferred over a narrow quarter that is clean in the twist. We are not talking about fat in the twist but muscling in the round. The more expensive cuts of meat come from the rear quarter."[29]

Forgason pointed out that his concept of an ideal breeding program also called for an animal "with a good spring of ribs and ample middle or barrel, one that is not waisty and is not pencil bellied either. We feel that a ruminant was designed to convert roughage into meat and milk. The cow must have enough capacity to ingest enough energy to produce milk for her calf while maintaining her own body in a satisfactory condition for reproduction."[30] Forgason also underscored the importance of sound legs and feet as a definite factor in longevity determined in part by the locomotion needed for daily rustling.[31]

Their experience has also shown that feet and legs are related to conception rates. Bulls tending to be post-legged, for instance, are predisposed to arthritis in their hips, stifles and hocks - making mounting more difficult. It has also been shown that weak pasterns will cause problems that result in lowered herd fertility rates.[32]

There has also been a concern regarding the length of the sheath on bulls and the length of teats on cows. "A bull with a long or heavy sheath, besides being prone to injury, is usually not as active sexually."* It has also been observed that such bulls will tend to sire females with such undesirable traits as long navel flaps, loosely attached udders and longer teats. And, of course, such females will also tend to produce long-sheathed bulls.[33]

As regards to size, some American Brahman breeders have selected against smallness; however, bigness has not been accepted or valued as a single important trait. In the Hudgins program a medium sized animal has been the ideal - 1,000 to 1,200 pounds for cows and 1,700 to 2,000 pounds for bulls, in pasture

*When British cattle were brought to Florida during the English occupation (1763-1783), it was found that few survived on the woodland pastures. Only the Spanish breeds continued to thrive. One theory suggested for the failure of the British breeds to propagate was that their relatively heavy sheaths were constantly irritated and scarred by dense palmetto growth - thus reducing fertility rates. (Jabo Gordon, Information Officer, Florida Department of Agriculture and Consumer Services, Tallahassee: October, 1975.)

condition. Indeed, such qualities as anatomical form, disposition and mothering ability have received first priorities.[35]

One of the highest priorities in the leading American Brahman herds has been cattle with good dispositions. Despite the lack of justification, "disposition is considered to be the greatest stigma of the breed."[36] Many breeders have worked hard with considerable success to erase this stigma down through the years. After the 1924 Importation, enough *Bos indicus* blood made it possible to cull most Brahmans that demonstrated uneven dispositions. In time, gentleness became a hallmark of the Hudgins' herds as well as many other herds. Fortunately this trait has been widely disseminated. Cattle buyers and breeders, who were once skeptics, have begun to realize just how really gentle Brahmans can be.[37]

As any successful cattle breeder will attest, careful record keeping is the cornerstone of any successful breeding program -a practice the Hudgins Ranch operation has sought to maintain since its earliest years. Histories have been kept on each animal sold with records back to both the sire and the dam. Periodic summaries and the analyses of the sire's records have also made it possible to tell which bull is doing the best job as far as economic traits are concerned.

Dr. Forgason recently pointed out about this particular breeding program:

> Animal breeding is not an in and out sort of thing, and you cannot change direction with each trend. When you do, you sacrifice prepotency, and prepotency is a very important economic aspect of our cattle. Type should be dictated by production, and production must result in a product that is acceptable to the consumer.[38]

The philosophies of breeding programs have varied considerably through the years and, of course, vary even today. Some Brahman breeders, for instance, are focusing more now

Mr. 3X Mucho Mucho 246/3, **owned by Joe and Emily Dial of Placedo, Texas, is a good example of the "growthy type" American Brahman. This particular animal won 30 championships in 13 months in five different states and sired a son which weighed 1,975 pounds by the time it was two years old.**

on producing larger sized animals - "size that leads to more pounds" as one breeder recently said.[39] And even with the highly successful Hudgins' breeding program, there has been some alteration in body length over the past few decades; however, their concept of an ideal weight has remained essentially the same.[40]

While there have been, and are today, many successful Brahman breeding programs, perhaps none has better demonstrated the importance of prepotency as an economic factor or the method by which prepotency can be maintained than the Hudgins' system of line breeding. But similar to all sound

The success of the Hudgins' breeding program depended so much on *Manso* that after his demise, his head was mounted in the Hudgins' office at Hungerford, Texas. This picture of Edgar Hudgins and the famous *Manso* was made by *Life* magazine in 1947.

breeding programs, this system has consistently illustrated the importance of considering the total animal. Regardless of one's concern for size or scale, what breeder can deny the importance of structural soundness, mothering ability, disposition, environmental adaptation, and all of the other characteristics that have made the American Brahman such an important breed.

Since the advent of *Manso*, over 30,000 head of *Manso* type Brahmans have been registered through the ABBA by the Hudgins Ranch while *Manso* lined Brahmans have won an overwhelming percent of championships in the show rings through the years.[41]

Other Renowned Bloodlines from the Ruffier Importation

While *Manso*, without question, has been the most famous of *Aristocrata's* sons, another celebrated male progeny of his was *Albacrata*, PH 730, the sire of *Albacrata, Jr.*, 37. *Albacrata, Jr.*'s blood can be found in many of the U.S. Sugar Corporation's top producers and show cattle. "It is interesting to note that the Brahman sire with the most Register of Renown points, *Sugarland's Rexcrata I*, PH 229/3, ABBA 160518, often referred to as simply *Sugar*, or as *'ole 229*, has a pedigree that combines the blood of *Aristocrata's* two best sons, namely *Manso* and *Albacrata*." Other important bulls of the Sugarland family descended from *Aristocrata* include *Rex Manso*, PH 426/2, who was National Champion in 1964, *Lox Manso*, *Albacrata Ponce*, *Loxacrata*, 1969 National Grand Champion and *Sugarland's Poncrata*, PH 4-25/5, 1972 National Grand Champion.[42]

According to longtime ABBA Director and lifetime cattleman James D. Sartwelle, "The best known son of *Aristocrata* used by the Sartwelle Brothers was *Aristarchus*, PH 681, ABBA 3018, nicknamed *Old Pard*, out of the *Imperator* daughter, *Faynette*, ABBA 8393. He was loaned as a two-year-old to Walter

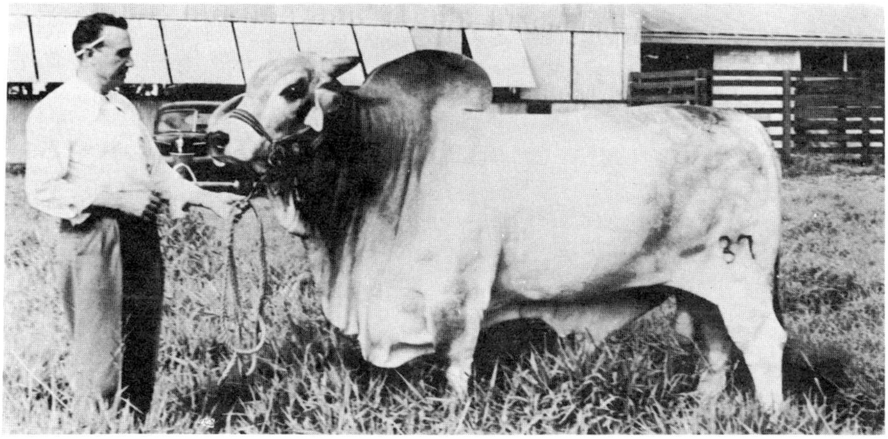

Albacrata Jr. 37 has dominated much of U.S. Sugar Corporation's pedigrees. He is being held here by Sid Crochet of U.S. Sugar at Clewiston, Florida. *Albacrata Jr. 37* was purchased by U.S. Sugar from A.J. Marceaux in 1948 for "the then unheard price of $15,000."

Aristarchus, PH 681, ABBA 3018, was the son of the Ruffier import *Aristocrata.* Born in 1936, he was 12 years old when this picture was taken on the Sartwelle Brothers Ranch.

Aubrey J. Marceaux of Kaplan, Louisiana, has been one of the pioneers in establishing the American Brahman as a principal breed in the great "bayou state". He was particularly instrumental in expanding the *Imperator* family of cattle within the breed beginning in the 1930's.

Hudgins. Hudgins also used several other sons of *Aristocrata* including *Prince Herosia*, PH 671, ABBA 3016; *Sir Anthony*, PH 733, ABBA 3029, and *Aristocrata's Prince*, PH 500, ABBA 3001, out of the *Alabama cow* (full brother to *Albacrata*)." A full sister to *Manso, Aristo Penine,* PH 950, ABBA 8704, was given to Walter Hudgins in June of 1938.[43]

Imperator, who as previously mentioned, was also first purchased by J. W. Sartwelle from Ruffier. "This animal had a short but productive life as he sired only five calf crops beginning in 1928. He made his mark through several outstanding sons and daughters through the skillful breeding of several breeders attracted to him."[44]

Among *Imperator's* notable sons were *Imperator II*, PH 327, ABBA 3001, and *Emperor of India*, PH 380, ABBA 3005, both of which were used as breeding stock by the Sartwelle Brothers. Another important son was *King of Kaplan*, who was used by Aubrey Marceaux of Kaplan, Louisiana, a former Director of the ABBA. *King of Kaplan* was half-brother to *Albacrata* on the dam's side, both bulls being sons of the Sartwelle cow, *Alabama*, 208 538. *Alabama* was sired by *Stanley*, PH 9, ABBA 105 of *Crescent V*, descended from *Teddy Roosevelt* and *Tippo*, a 1906 bull. Among *King of Kaplan's* famous sons were *King of Kaplan, Jr.*, *Imperator Manso*, PH 699, ABBA 2088, and *King Grey*, PH 230, ABBA 3561.

In the 1950's, Marceaux sold *King Grey* for $10,000.[45]

Several daughters of *Imperator II* were used with sons and grandsons of *Imperator* to concentrate his blood. Among them was *Commander Rose*, PH 81, ABBA 13051, *Imperator II's* daughter, bred to *King of Kaplan* to produce *Imperator Manso*.[46]

Imparistre, out of an *Aristocrata* daughter was used by the late Henry O. Partin of Kissimmee, Florida, also former president and director of the ABBA. One of the calves owned by Partin which was sired by *Imparistre* was sold to the Norris Cattle Company of Ocala, Florida, for $5,000.[47]*

Three notable grandsons of *Imperator*, sired by *Imperator II* and used by the Hudgins Ranch were *Imperator Ulfilas*, PH 747,

*Norris Cattle Company, a big name in the cattle business for many years, kept approximately 500 breeding Brahmans and sold Brahmans to many parts of the world. In fact, according to Byron Johnson, Norris' former purebred herd manager, Norris helped to pioneer the aerial shipment of cattle in 1946.

In 1980, Norris Cattle Company liquidated their Florida cattle operation selling the best part of their herd to Mann Bailey of Oxford, Florida. However, the company still maintains cattle interests in South Africa and Zimbabwe. At one time they maintained partnerships in American Brahman operations in Costa Rica, Argentina, Ecuador, Mozambique, South Africa, Rhodesia and the Philippine Islands.

Two old timers, Florida cowman Henry O. Partin and one of his outstanding herd sires are pictured here at the Partin Ranch near Kissimmee, Florida. Partin and his family did a great deal to popularize Brahmans in the Southeast.

ABBA 3036, later sold to Vernon Frost; *Rexcardo* PH 759, ABBA 3047, also used by O. F. Spear, Sr. of San Antonio and J. V. Gates of Poteet, Texas; and *Commander Van Dorn*, PH 829, ABBA 4381, used by the Hudgins from 1940 to 1950.[49]

One of the most celebrated descendants of the Sartwelle Brothers' *Imperator* was a grandson named *Emperor*, PH 200, ABBA 7961. Henry O. Partin, with the help of *Emperor*, practically transformed Florida's cattle industry during the 1940's and 1950's.[50]

Domesticated cattle had been brought to Florida as early as 1521 by Ponce de Leon, and the cattle industry was indeed Florida's oldest industry. However, the tough, scrubby cracker cattle, descended principally from early Spanish herds, had ranged over Florida's river plains and palmetto-pine forests for over 300 years, and they had changed very little. They were far from being ideal beef cattle. Some attempts to upgrade native stock with British breeds and even a few *Bos indicus* cattle had been partially successful, but until Partin, and then a few other Florida ranchers began to popularize the use of Brahmans, the general quality of Florida beef had been poor.[51]

In the 1930's, Partin purchased one of *Imperator's* sons, *Imparistre*, PH 423, ABBA 4435, from Pat Johnston of Orange County, Florida. *Imparistre* was bred with the best Brahman cow which Partin had. This union produced Florida's first Brahman prodigy, *Emperor.*[52]

Never defeated in a show ring, *Emperor's* fame grew through the 'forties and he was used as a sire by a number of Florida's cattlemen, including important breeders A. Duda and Sons, whose *Jesse de Emperor* was the Grand Champion bull at the Kissimmee Brahman Show in 1957. A few years earlier Partin's Brahmans, all descended from *Emperor,* won 108 blue ribbons in a two-year period in shows around the nation.[53]

Because of his beef-like qualities and excellent progeny, *Emperor* was leased for six months in 1946 by Edgar Hudgins of the J.D. Hudgins Ranch, and he was bred to many of Hudgins' top two-year-old heifers. Thirty-seven offspring resulted. Thus, the great Florida-bred herd sire had an opportunity to visit the state where his grandfather, *Imperator*, had immigrated in 1924. In was probably the first time in the breed's history that a Southeastern breeder had loaned stock to a Texas breeder.[54]

The *Imperator* blood has also been important in the development of the W. H. Stuart Brahman herd at Bartow, Florida and the Sugarland Brahman herd at Clewiston, Florida. Both the Stuart and Sugarland Ranches have been extremely important

J. D. HUDGINS, INC.

BREEDER OF

BEEF-TYPE BRAHMAN CATTLE

HUNGERFORD, TEXAS
December 25, 1946

Mr. Henry O. Partin
Kissimmee, Florida

My Dear Old Timer:

This being Christmas day, I have taken a few minutes off and will tell you something of my stay in Texas up until now. Today's dinner will be something special for you and all of the Partins but with me it is Christmas dinner all of the time. My wives and I are doing fine and I have gained over two hundred pounds since I got here.

Boy, was I upset when a red truck, driven by a big slow moving fellow, loaded me off of my island and drove away. I thought that my old friend has treated me mighty bad and thrown me away. I thought that he had forgotten all of the good times that we had had together. The times like when I had won the big Florida Brahman shows and all eyes were on me and how you smiled and slid your hat back on your head a little and went walking around on air.

Well we rode for two days and stopped once and picked up an old bull named King of Kaplan. They all said that he was my uncle. I would have gone to the mat with him when they first put him in the truck but the old slow moving fellow had a steel gate between us and we couldn't get together. I gave him a good cussing tho and watched him all of the way to Texas.

We finally got to a little stop on the highway and the old slow moving fellow got out and told his wife that we were home. It was a little before dark and in a few minutes there was a crowd of folks around. I felt a little better then they started to saying so many nice things about me, but I kept looking for a Partin and thinking about my island I worried so on my trip over that I did not eat or drink very much. YOu know how you feel when you get away from Florida and start thinking about home.

They finally got around to getting me out of the truck and I was sure glad to stretch my legs and feel the ground again, but it wasn't me Florida ground. They didn't give me much to eat that night but the next morning I got some real food and have been doing fine ever since.

About noon of that same day, they opened a door and, old friend, I knew right then you had not let me down. There were six of the most beautiful cows eating grass about a hundred yards away that I most ever saw. Boy, them women were fat. I didn't walk, I troted over to get acquainted with them. At first, they were bashful and a little scared of me but I knew that would be only for a few days. Well, to make a long story short, I married all six of them in less time then three weeks. These women, I learned later, had all kinds of show records and I know that you will be proud of my sons and daughters.

After I had been in Texas about a month, and about to forget my island, they put me with fifty heifers that were a sight for sore eyes. These young gals wouldn't let me get close to them at all for a while. They were scared to death of a man. I later learned they had never been with any men before and such a proud fellow as me would naturally scare them. Well, since the day I was out, I have married quite a number of them. I have been so busy I have forgot the number. I have also forgiven you for sending me away from my island. I hear that I will be going back to Florida in the spring and naturally I will be glad but if I could have my wish I would be two bulls-one of me in Florida and the other in Texas.

I get a kick out of cussing a whole herd of big Brahman bulls that are close by. Since there is a double fence between us, I call them all kinds of low down names and threaten their lives. I get my horns all muddy by hooking in the dirt and I walk the fence and cuss them some more and all of the time my wives are eyeing me proudly. Those bulls don't know it, but I would run like the very devil if that crew could get to me.

I better get back to my wives but I want to wish you a merry Christmas and a very happy New Year and tell you that I have forgotten all of the bad things that I thought about you when you sent me away.

EMPEROR

Edgar Hudgins, of J.D. Hudgins, Inc., wrote this humorous letter to cattleman Henry O. Partin in 1946 after Partin's great bull *Emperor* had been brought to Texas. *Emperor* was used as a herd sire on the Hudgins Ranch for six months and was bred to many top, two-year-old heifers.

in the development and the promotion of the American Brahman in Florida and other areas of the Southeast.[55]

Another little known Brazilian bull that left his imprint on the development of the American Brahman in Florida was an excellent Guzerat bull named *Tartaruga*. This bull was brought to the Hudgins Ranch in 1924 along with the 30 other Brazilian bulls that reached that ranch from Mexico. Walter Hudgins sold *Tartaruga* to his cousin Joel Hudgins of Wharton, Texas, who also had a well-bred Brahman herd at the time. Many excellent daughters resulted from the use of this bull. Hudgins named them *Miss Texas First, Miss Texas Second, Miss Texas Third*, etc.

When Joel Hudgins died, his family put his cattle up for sale. Hearing of the sale through Walter Hudgins, Henry Partin purchased 125 of these top cattle and shipped them to Kissimmee. It was *Miss Texas Forty-fourth* that Partin later mated to *Imparistre* producing the great *Emperor*.[56]

One of the bulls in the 1924 Importation that was never registered, but that still made an important contribution to the Brahman industry was named *Dick* (not be be confused with *Dick Brownie*, ABBA 968, or *Dick*, ABBA 4135). *Dick* was a light blue bull described by Dr. W. S. Jacobs as an Indu-Brazil type.* He was sold directly to Bassett-Blakely. After being used on Bassett-Blakely's fine herd, *Dick* and the same herd were sold to Joe D. Hughes. *Dick* was later purchased by Dr. Jacobs and continued to make the rounds, as so many of these imported bulls did, and was eventually loaned to the Pierce Estate. One of the most important bulls descended from *Dick* was *Red Dick*, ABBA 2574, owned at one time by Milby Butler.

Another important bull brought to Texas during the 1924 Ruffier Importation was *Conquista* who was brought to Artesian Wells, Texas, by John Martin, later becoming a valuable sire for cattleman Calcutta Kimball.[57]

Santinho, ABBA 43, was bought from John Martin by cattleman Jake Reitz who later sold him to Dr. Jacobs. After being

*There were supposedly no Indu-Brazil cattle in the 1924 Importation, but Dr. Jacobs obviously made this comparison later after he had seen *Indu-Brazil* cattle, perhaps, in the mid-thirties.

"Tartaruga"

SOCIEDADE DO HERD BOOK
ZEBÚ
UBERABA

U. S. DEPARTMENT OF AGRICULTURE
Certificate Issued
☆ MAY 26 1924 ☆
L. W. D. For Chief of Bureau.
BUREAU OF ANIMAL INDUSTRY.

Certificado de Origem 6L

A Sociedade do Herd-Book Zebú certifica que o reproductor abaixo descriminado do sexo masculino é de puro sangue da raça Guzerat sendo filho do touro puro Guzerat "Thesouro" e da vacca pura Guzerat - Nubia

DESCRIPÇÃO:

Nº da chapa:
017

Nº da inscripção no Herd Book Zebú:
Registro geral 017

Marca:
ZB - HZ

Signaes:
ʒ

Côr:
Azulego

Nasceu:
9 - 9 - 1920

Observaçoes:
Vaccinado contra o Carbunculo symptomatico. Immunizado contra a Piroplasmose Bovina - (Tristeza)

Uberaba, 30 de Abril de 1923
Minas Geraes (Brasil)

pela Sociedade do Herd Book Zebú

Gualdino Rodrigues da Cunha presidente
José Affonso Ratto 1º secretario

The *Sociedade do Herd Book* of Uberaba, Brazil issued registration papers for the bulls that were in the Ruffier Importation. This is the certification for the bull *Tartaruga* eventually purchased by Joel Hudgins of Wharton, Texas.

used for several seasons by the Hudgins Ranch, *Santinho* was sold to the House Brothers of Hockley, Texas.[58]

Still another 1924 bull, *Amador*, ABBA 237, was owned by the Hudgins Ranch, and was described by Dr. Jacobs as "a very good bull." Another bull, *Dan*, later known as *Champion Dan*, was the sire of *Dick*, ABBA 1422. *Dan* was owned by the Stark Estate of Orange, Texas. He had the important honor of being the Champion Brahman bull of the Houston Fat Stock Show in 1932.[59]

Dick, on the other hand, was later sold to Coon and Culberson and sent up to Dalhart in North Texas. There he was pastured with 200 Brahman cows. "The Brahman calves that resulted from that mating were topped and brought to the Houston Fat Stock Show and made a marvelous exhibit by Coon and Culberson, but as they were unregistered, they were not used in the competition . . . ." These offspring of *Dick* were eventually sold to Milan G. Simon of Port Lavaca, Texas, and in 1941, they were registered in the ABBA by a committee of five consisting of Walter Hudgins, Dr. W. S. Jacobs, J. W. Sartwelle, Marshall Johnson and Fred Neuhaus.[60]

These cattle were eventually sold to Frank Buffaloe and C. M. Frost with Buffaloe receiving most of them. Buffaloe's herd contributed in large measure to the development of the Whitaker family herd with the Whitaker herd sire, *Sugarland's Rexcrata, I*, producing some of the better known progeny when bred with Buffaloe cows.[61]

Jacobs later said, "These cows by . . . *Dick*, son of *Champion Dan*, have made a startling impression upon the lovers of fine Brahman cattle wherever they have been shown." *Dickson #4*, ABBA 8690, and *Dickson 1A*, ABBA 8691, were Raymond Dickson's choice among the Ruffier Importation and both were also considered excellent bulls.[62]

Another of the little known 1924 Brazilian bulls was *Jararaca*, a large, dark grey bull. Bought by J. D. Hudgins in May of 1927 from John Martin, this sire left some outstanding daughters in the Will Border herd at Hungerford, Texas. "The best individuals were later purchased by the Hudgins Ranch."[63]

J.V. Gates of Poteet, Texas, used the bloodlines of *Imperator* **and** *Aristocrata* **in the 1940's to produce what he phrased "The Gentle Kind." Pictured above is the first sire,** *Sir Manso,* **to be used in his program.**

Using the bloodlines of *Imperator* and *Aristocrata*, J. V. Gates of Poteet, Texas, started in the Brahman business in the 1940's. His early goal was to produce an animal that would be naturally gentle, "not an animal that had to be tamed, but one with a bred-in gentle disposition. . . ."[64]

Starting with *Sir Manso*, sired by *Manso*, and eleven cows Gates had purchased from Walter Hudgins, he also added heifers to his herd from the Sartwelle Brothers Ranch. Breeding an *Imperator* sired heifer with *Sir Manso*, *Barzee's Perfecto No. 53* was produced - what Gates called ". . . the nearest to a perfect bull for what I wanted. . . ." Although *Barzee's Perfecto, No. 53* died soon after, he left an impressive legacy of 100 daughters and several good sons.

Gates had taken the name *Barzee* from the brand *Bar Z* that he had especially selected for his own Brahmans. In time, the name *Barzee* became synonymous with *Barzee's Perfecto, No. 53*. By the 1950's Gates' *Barzee Brahmans* were making a mark on the show circuit as "The Gentle Kind."[65]

Gentleness, of course, has been emphasized by most progressive Brahman breeders. And no one has done more to develop and promote this important trait than cattleman J. T. White of Hearne, Texas. White, whose family has ranched in Texas since 1822, was one of the most popular of the ABBA presidents.

White has always considered his Brahmans as a part of his family. He and his wife Katie have spent many hours making "pets" out of their small herd of purebred Brahmans, and it has paid good dividends.

This affable cowman has stayed up many a night to help a heifer deliver a calf or to keep a newborn calf warm and alive during a cold northerner. He has his own technique of "delivering" calves and maintains what he refers to as "three sets of maternity wards."[66]

Both the Whites believe in the returns of hard, honest work, and through the years they have hired many young people to help on their ranch. A good part of their profits from their cattle operation has been used to pay for scholarships for needy, deserving students at Texas A&M.

During the early 1950's, J. T. White had the opportunity of addressing a joint session of the U. S. Congress regarding a serious drought that had practically paralyzed the agricultural industry in Texas. Texan Sam Rayburn, then Speaker of the U. S. House of Representatives, said that he wanted Congress to hear about the rancher's problems "in the venacular of an old cowhand talking to some other cowhands. . . ." White used no notes in his speech, but with his usual warm, persuasive and natural delivery, he won an enthusiastic reception from the national lawmakers.[67]

Another Texas cattleman known for his gentle Brahman cattle has been J. Carter Thomas of Cuero. Thomas has also been a successful dairy cow breeder, and by crossing American

J.T. White (ABBA President 1955-1958), from Hearne, Texas, is one of many Brahman breeders that has helped dispel the misconception that Brahman cattle have a bad disposition. "There is not a more gentle animal alive," according to lifetime catttleman White.

Brahmans with Jerseys and Brown Swiss, he has developed a derivative dairy cow known as the "Milk Master."[68]

The W. H. Stuart Ranch of Bartow, Florida is one of many Florida Brahman breeders that has also stressed the importance of disposition, and their bulls have been noted for their good temperament.[69]

Dr. A. O. Rhoad, one time head of the New Iberia Livestock Experiment Station in Louisiana, who was also closely involved

Most of these top Brahmans at the Partin's Heart Bar Ranch near Kissimmee, Florida, are descended from the *Imperator* bloodline.

in investigative work on the Brahman breed both in Latin America and Louisiana, and who completed work on the Santa Gertrudis, commented in 1946 on the importance of the Ruffier Importation:

> Before the 1924 Importation, one must admit that no great progress was made in improving the then rather ordinary types. The "Brahman" had not "taken hold" in a general way, only a few devoted breeders then existed, and the number of Brahman cattle was not great. Following the 1924 importation, increased interest developed due in large part to the excellent quality of a number of imported animals. They were large, strong boned, and meaty individuals albeit of mixed Indian breeding of predominately Guzerat blood with evidence of some Gir and Nellore.[70]

While Dr. Rhoad emphasized that the quantity and quality of the animals in the 1924 Importation "increased interest" among breeders, it should also be pointed out that the willingness coastal breeders showed toward loaning, leasing, and selling these bulls to other breeders helped immeasureably in improving and popularizing the Brahman. The spirit - with most breeders at least - must have been almost evangelical in nature. Those who had had the opportunity or foresight to buy these fine new bulls were obviously proud of their investments, but they also seemed to feel a commitment and a desire to share their "new discoveries" with other breeders. There was an interest on their part not only to improve the Brahman breed but even to share in the pleasure of doing it.

The Morias Importation

A year had not passed before a Brazilian cattleman, J. Morias, shipped 120 more Zebu bulls and 18 heifers into Mexico from Brazil. Landing on the West Coast of Mexico, Morias intended to display his herd at the West Coast Exposition of Mexico; however, his plans were blocked when a revolution suddenly broke out. Cut off from his sponsors in Brazil and low on money, Morias gruelingly trailed his herd by foot across Northern Mexico's rugged interior. Managing somehow to keep his herd intact, he turned them out on the Centinella Ranch near Piedras Negras. Morias hoped to bring his cattle across the Rio Grande at this point into Eagle Pass, Texas.

Frustrated by financial problems, Morias turned to American engineer and banker, William A. Smith, for help. At the time, Smith was engaged in building a railroad in Mexico. Although not a cattleman, Smith knew of the growing demand for

Brahman cattle in Texas, and he took a contract to buy all of the heifers and to loan Morias money, using certain of the Brazilian bulls for collateral. Smith planned to bring the Brahmans across the river and up to his father-in-law's ranch at Victoria, Texas. However, his father-in-law, James Welder, refused to allow him to pasture the cattle there because of the recent outbreak of aftosa. Desperate now for pasture, Smith contacted Dr. W.S. Jacobs, "as the only sucker he could find" Jacobs later mused.[71]

In November, Jacobs went to Mexico where he purchased all of the 18 heifers and 6 of the bulls that Morias was holding. During the next spring, the cattle he had purchased were cleared to come across the bridge at Eagle Pass. From there they were shipped up to Jacob's ranch at Webster, and before the year was out, Morias, Smith and Jacobs had transferred the rest of the bulls to Webster.

The Brazilian bulls brought in at this time were fairly well dispersed except, of course, for those that were destroyed when aftosa broke out again on the Jacobs ranch. However, Jacobs was able to register the heifers and their offspring in the ABBA Herd Book with an "AA" classification, and maintained a monopoly on these bloodlines until the late 1940's.[72]

Some of the original buyers of this 1925 Importation were Jake Reitz, Tom East, Chub Poole (Hogue Poole's father), the Sartwelle Brothers Ranch, Henry Kimball, Hogg and Dickson of Nueva Laredo (who kept six bulls in Mexico and later sold them to the Pierce Estate), Ed Lasater, William Carroll and Jess Stewart, Rucie A. Moore of Orange, Texas and Vinton, Louisiana, and Fred Neuhaus.[73]

Another outbreak of aftosa was confirmed by the State Livestock and Sanitary board and the U.S.D.A. at the ranch of Dr. Jacobs on July 29. "The infected herd of 96 cattle was slaughtered and burned on July 31," and a quarantine was again imposed on parts of Harris County and later extended into portions of Galveston and Brazoria Counties. Although it was presumed that this second outbreak of aftosa since September of 1924 was caused by the Morias Importation, it was never proven. Several theories regarding the "mechanical transfer" of this disease were suggested. One of the most popular was that the "aftosa germ" had been transferred on plants that the Brahman cattle ate. Allegedly these plants were brought into Texas by Japanese nurserymen. At any rate, the last infection was found on October 11, and the quarantine was lifted some months later.[74]

Dr. Jacobs had a fondness for naming cattle after certain people he knew and respected, and he named one of the best bulls that was imported in 1925, *Red Imes*, after Dr. Marion Imes of the U.S. Bureau of Animal Industry. Dr. Imes had been put in charge of controlling and eradicating the 1924 Foot and Mouth disease, and he had supervised the operation in such a tactful and professional way that he won the "high esteem" of Jacobs and others that were directly involved.[75]

The history of *Red Imes*, a Gir type bull, is an interesting but short one. He and the heifers he was penned with at Dr. Jacobs' ranch were all slaughtered shortly after the 1925 Importation. And consequently, no progeny resulted from the mating that took place on Texas soil. However, during the six months of quarantine at Piedras Negras, *Red Imes* was exposed to all 18 of the Brazilian heifers that Jacobs purchased. "From that exposure there resulted 7 calves." The first of these born was named *Jimmy* after Jacobs' friend, James W. Sartwelle. *Jimmy* was sold to Walter Williams, who later exported him to Cuba. As he matured, *Jimmy* ". . . made a notable record in the annals of Brahman cattle in Cuba."[76]

Of the heifers that Dr. Jacobs bought from this importation, *Hattie Cleo (Old Bob)*, ABBA 1409, was probably the most famous. From *Hattie Cleo* came *Henrietta*, ABBA 1413. Other Brahman calves born of the 1925 imported heifers included *Ruby*, ABBA 1412, and *Mary Jane*, ABBA 1414. "These were all," according to Jacobs, "pure Guzerat cows."[77]

Red Imes was never registered by ABBA but did have papers issued by the Sociedade do Herd Book Zebu of Uberaba, Brazil. This bull was evidently sometimes confused with another solid red bull that Jacobs, for obvious reasons, had named *Solid Red*. Also an imported bull, *Solid Red* was sold to Milby Butler and used one season by Jake Reitz. Butler later wrote of *Solid Red*, "With this bull . . . I did a considerable severe line breeding. . . . this line breeding brought some very outstanding animals and some very sorry ones." One of *Solid Red's* important offspring was *Estrelo*, ABBA 287. Butler rented *Estrelo* from Jacobs for awhile and later wrote, "This . . . bull I think was the best bull of all the animals imported in this country that I have seen."[78]

Another bull from this importation bought by Jacobs was a dark bull with an excellent conformation named *Normando*. Jacobs bought him directly from Morias, used him for one year and sold him to Jess McNeel of Southeast Texas. McNeel sold *Normando* to Dr. J. L. Ducroz of Brazoria County, Texas.

Tango, another bull purchased by Jacobs from Morias, "was a very handsome bull . . . that carried with him a certificate of Grand Champion at the Rio de Janerio Brahman Show of about 1922 or 1923." It was estimated that he weighed in excess of 2,500 pounds and was described by Abel Borden "to be a good specimen of the Krishna Valley breed." *Tango* actually serviced one of Jacobs' cows while he was still in Mexico. The offspring of this union was *Tangarina*, ABBA 1412. When *Tango* was brought to Dr. Jacobs' ranch, he was exposed to foot and mouth disease and unfortunately was one of many that had to be destroyed.[79]

Cornelius, ABBA 4466, was imported by Morias and Jacobs but was among the bulls bought by W. D. Cornelius and Johnny Walker. Interestingly, he was also used by Victor LaTulle of Bay City, Texas, in creating his own Braford herd. After this, *Cornelius* passed into the hands of Stewart and Francis Savage. "This bull was used by Savage for two or three years and was a wonderful producer."[80]

Estrella, ABBA 242, was an Indu-Brazil type bull that was imported during the Morias Importation along with a red bull named *Rochinal*, ABBA 181. *Rochinal* was purchased by the House Brothers while *Estrella* was loaned (presumedly by Jacobs), to Milby Butler of League City, Texas, and later sold to Joe D. Hughes who used him with his red Brahmans, "part of which

Two of the most revered pioneer Brahman breeders were Will Cornelius of Matagorda County, Texas and Henry O. Partin of Osceola County, Florida.

he [Hughes] later sold to Walter Fondren of Houston." *Estrella* was also the sire of the Brahman bull *Ont*, ABBA 5853. Another of the bulls that came in the 1925 Importation was *Faceiro* ABBA 6227. L. P. Bunge of Garwood, Texas, used *Faceiro* with his Hahn cattle herd. "The resultant progeny became the foundation of many of today's red herds." (See Chapter V, "A Preference for Reds.")[81]

Though never so important as the Ruffier Importation in the development of the American Brahman, the 1925 Morias Importation had its greatest effect on Texas herds. But some Brahman breeders in Louisiana and Cuba eventually received some of these cattle and their offspring. Dr. W.S. Jacobs was certainly affected most immediately, and it was largely through the *Bos indicus* heifers he imported, and their offspring, that he established the foundation of his so-called "famous 100% imported Brahman blood herd," which were classified as "AA" by the ABBA until 1941. "The original Jacobs' brand was , and when the ABBA eliminated all class distinctions, Dr. Jacobs branded AA on his cattle so the double 'A' appeared on the registration papers only as a holding brand."*[82] But when the "AA" and "A" was discontinued by ABBA, Dr. Jacobs still continued to brand all his cattle from this bloodline with "AA." Even when this bloodline passed on to Jacobs' son, the "AA" brand was continued, and when these cattle were sold, contracts specified that "when they were bred together they must carry an "AA" brand." There are some herds that carry these brands even today.[83]

Other *Bos indicus* cattle which entered the United States during the Ruffier and Morias Importations included *Goliath*, ABBA 45; *Vatapa*, ABBA 190; *Selim*, ABBA 233; *Sambogni*, ABBA 239; *Laura*, ABBA 1399; *Lucile*, ABBA 1400; *Coley*, ABBA 1401; *Pony*, ABBA 1403; *Duchess*, ABBA 1404; *Daisy*, ABBA 1402; *Ethel*, ABBA 1408; *Alice B*, ABBA 1411, and *Tanico*, ABBA 394.[84]

*The "holding brand" was a part of the private herd number.

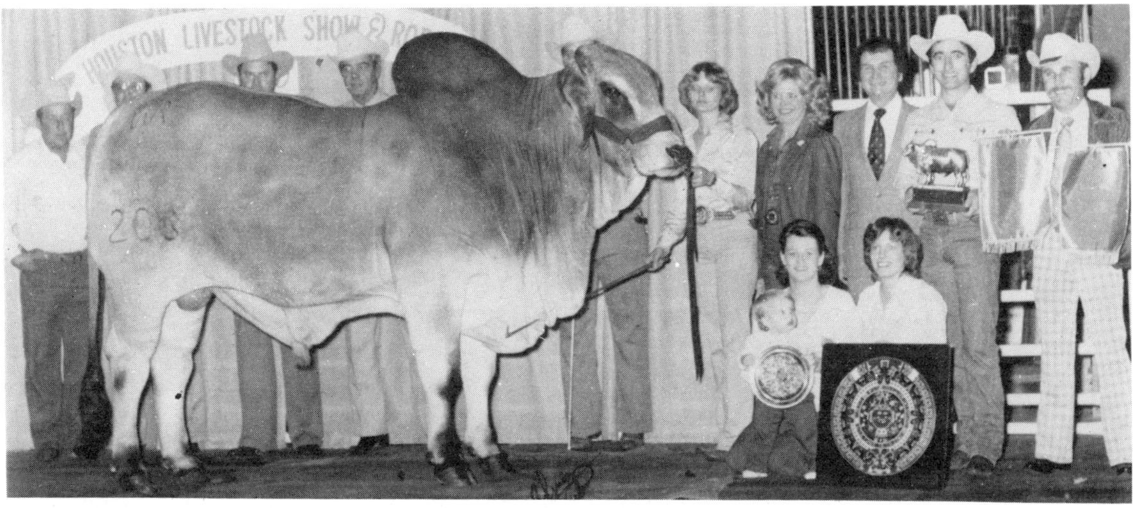

The great Brahman sire *W. R. Suva 203* is in large measure the modern product of the Brazilian imports of 1924 and 1925, which includes some of the best bloodlines from the Koontz Ranch of Inez, Texas. This bull is jointly owned by the Watts Ranch of Frost, Texas; the V8 Ranch of Hungerford, Texas; and Charlie Thomas Brahmans of Houston, Texas. During the 1979-80 season, this bull won the Triple Crown Grand Championships at Dallas, San Antonio and Houston. (One-half interest in this sire was sold for $75,000 in 1979.) Pictured above with *W. R. Suva 203* are (left to right) Max Watts, Glen Urban, Judge Bill Turner, Bob Massey, Jim Williams, Tammie Watts, Mr. and Mrs. Charlie Thomas, and David Barnes. Seated are Mrs. Jim Williams and Rachel Williams and Debra Watts Barnes.

Related to many of the famous early Brazilian imports was *Suva 1st*, at one time the senior herd sire at the Koontz Ranch of Inez, Texas. (This ranch is now owned by the son-in-law and daughter of Henry Koontz, Mr. and Mrs. Robert L. Massey). In 1949, this bull was loaned to Texas A&M and was one of the first Brahmans ever to receive a Record of Performance Test - "graduating" in the top ten percent of all cattle in his class.

The Nunez Importation Into Mexico

Even before the 1924-25 Importations of Brazilian bulls into Texas, some *Bos indicus* cattle from Brazil had been imported into Mexico. In 1923, for instance, Pedro Marquex Nuñez directed the transfer of 82 head of Nellore and Guzerat type bulls from Brazil to the Mexican port of Tampico.

One of these 1923 bulls was a Guzerat named *Pontal* which was described as "the champion of the Guzerat race of Brazil in the year 1923." Another was a three-year-old Nellore type named *Satan* which had won the second prize in the same exposition held at Rio de Janerio, Brazil in 1923.*[85]

The first independent cattlemen to acquire any of these bulls were Señor Samuel Melo y Ostos of the Rancho De Chila Cortaza in Lantoyuca, Vera Cruz; Senor Antonio Chirinos of the Rancho de Las Piedras in Panuco, Vera Cruz, and Senor Manuel E. Guzman of El Rancho de Salinas, also in Panuco, Vera Cruz.

Other Mexican breeders who acquired Brahman bulls from this importation included Senor Amado Jauregui and Senor Andres Herrera from Vera Cruz; Don Eloy Gomez and Mr. Reader of Altamira, Tamaulipas; Senor Gonzalo N. Santos and Senor Luciano Era of Guerrero, San Luis Potosi.[86]

Manuel Guzman Willis wrote in 1945, "From these Brazilian animals, magnificant results were obtained in the region of Tampico and the Huastecas, from data sent from Tabasco, they also had good results."[87]

The 1946 Importation

A series of eccentric events was involved in the eventual entrance of 18 Brazilian *Bos indicus* bulls into Texas in 1946. According to the principal purchaser of these bulls, Senor Esteban Garcia, it all began as sort of a misunderstanding between Mex-

*The Pierce Estate earlier tried to buy *Satan* but because of quarantine problems was unable to work out a trade with Nunez.

ico's Secretary of Agriculture, Senor Marte Gomez, and a group of Brazilian cattlemen.

Gomez had made a trip to the Brazilian cattle capital of Uberaba to attend a cattlemen's banquet. When the banquet ended, Senor Gomez made a speech expressing his gratitute to his hosts, and made this statement, "I hope in the very near future we will be able to get some of these cattle to improve the cattle industry in our country." He evidently made this remark as a way of being nice rather than as a request for Brazilian cattle.[88]

Taking Gomez at his word, thirty Brazilian cattlemen got together, hired a boat, and soon had a load of approximately 120 bulls, mostly of the Indu-Brazil and Gir types, headed for Mexico. The Mexican government had not been advised of the shipping, but when the ship anchored off the Mexican coast, Gomez advised the Captain that Mexico had an agreement with "all these other countries prohibiting the importation of cattle from aftosa countries."[89]

According to the Garcia account, ". . . these Brazilian people came to the Brazilian Ambassador in Mexico City . . . . This was just after World War II, and Brazil had been an ally of the U.S. Then U.S. Ambassador Daniels was contacted . . . ." Daniels evidently contacted the U.S. State Department, and they acted to help the Brazilians, without consulting the U.S.D.A., by arranging with the Mexican government to place the animals on Sacrificios Island, near Vera Cruz. When these cattle were placed under strict quarantine, Dr. E. C. "Buck" Sharman of the U.S.D.A. was one of the U.S. Government inspectors sent to help inspect the cattle.[90]

Anxious to find buyers for the cattle, Secretary Gomez contacted the Garcia Brothers of McAllen, Texas. Esteban Garcia, who along with Eligio ran cattle on both sides of the Rio Grande on ranges that had been ranched by their families since 1740, agreed to come to Sacrificios Island and have a look.[91] Senor Manuel Guzman Willis of Tampico also examined these bulls while at Sacrificios Island, but he expressed no interest in purchasing them.[92]

Because of the large expense involved in maintaining the bulls during the quarantine, the Garcias also advanced the Brazilian interests "a large sum of money." For this favor they were given ". . . the privilege of selecting the best cattle . . . at a certain price," and he later contracted "to select the first 25 bulls in that shipment." For these, Garcia allegedly agreed to pay from $1,000 to $2,500 each. Senor Manuel Guzman Willis reportedly was asked $750 for each of these Brazilian bulls earlier, presumedly for the whole lot.[93]

When the 25 cattle were declared free of any infection, Garcia had them transferred to a Mexican ranch. As the bulls were greatly emaciated by this time, Garcia had great trouble in transferring them across the interior to his ranches. Shipping them by rail, he was impelled to make several stops along the way. In the state of Coahuilo, he was forced to borrow money from the government having to leave 7 bulls there as collateral. It was also necessary for him to leave several bulls on another ranch; consequently, when he reached home, he had only 13 bulls left. Four of these 13 later went to D. V. Guerra, his brother-in-law, and to his brother Eligio. These were all brought across the Joe Pate Bridge at Hidalgo, Texas, after having met all the health requirements imposed by the Texas Livestock Sanitary Commission (now the Texas Animal Health Commission) and the U.S.D.A.[94]

Although these imported bulls were described variously as "Girs, Indu-Brazil and even Nellore and Guzerat," Dr. James O. Sanders said, "In at least most cases, the actual breeding of the bulls was not known." While most were Gir and Indu-Brazil types, *Aimore*, ABBA 33877, was believed to be a Guzerat and *Carijo*, ABBA 32901, was thought to be a Nellore.[95]

The other five Brazilian bulls, making a total of 18, were presumedly purchased by Joe and Bill Pate and Hogue Poole.[96]

The Garcias eventually crossed their Brazilian bulls with cows from the Hudgins' herd, and as Esteban Garcia later reported, "We had great results." In time, the Garcias had between 400 to 450 mother cows scattered over their vast ranges.

A number of other cattle breeders, including Dr. Jacobs, the Clifford Brothers, C. E. Yoakam, and Jess McNeel, later borrowed or leased some of these bulls from the Garcias.[97]

The late C.E. Yoakam, founder of the Cherokee Ranch at Cherokee, Texas, with Mike Clifford and Dr. C. C. Conley at 1954 Houston Fat Stock Show. Yoakam, former ABBA Director, did much to popularize red Brahmans. Mike Clifford and his brother J.A. Clifford were pioneer red Brahman breeders in the Texas Valley.

Of the bulls in the 1946 Importation, Esteban Garcia later said that he thought *Repucho*, ABBA 33875, was the best of all. *Repucho*, "a gray Indu-Brazil bull with extremely good breeding characteristics," was used as a herd sire by the Garcias as well as by such breeders as C. E. Yoakam and Dr. Jacobs. Of the imported red Brahmans, Garcia said that he felt *Rio Negro*, ABBA

33876, was the superior bull. "*Rio Negro* possessed extremely good Gir features and a very smooth conformation." The Clifford Brothers later developed a herd of red Brahmans with a very high percentage of *Rio Negro* blood.

Arauto, ABBA 32904, owned by Joe and Bill Pate, was generally considered to be the nearest to a pure Gir in the 1946 Importation. He was used by the W. W. Fondren Estate and later by C. E. Yoakam. "He has been considered to be the progenitor of many of today's American red Brahmans." The largest framed bull and probably the only Guzerat of this importation was *Aimore*, ABBA 33877, used by Dr. Jacobs.[98]

Other important bulls in this importation included *Brilhante*, ABBA 33878, "a medium gray Indu-Brazil who sired many red calves; *Gaucho*, ABBA 33871, a red dappled Indu-Brazil, sired *Indo-Gaucho*, among many others. *Indo-Gaucho* became one of the Kimball Ranch's best red Brahman herd sires"; *Precioso*, ABBA 33870, "a medium gray Indu-Brazil also was used in many red Brahman breeding programs." *Nube Gris*, who stayed in Mexico, was an important sire in "the development of the Reyes Garcia Indu-Brazil cattle in Mexico"; *Jacarandar*, ABBA 34089, "an Indu-Brazil purchased by Hogue Poole," has been an important sire for those breeders desiring to emphasize Indu-Brazil features.[99]

Certainly those Brahman breeders who were particularly interested in developing red Brahman herds received a great boost from the 1946 Importation, but as Joe Dial of the 3X Ranch at Placedo, Texas pointed out, "the effect these bulls had on the development of both red and grey Brahmans registered in ABBA has been noteworthy."

Perhaps no American breeder utilized the progeny from this importation more than C. E. Yoakam. At one time, his "herd sire battery was composed of 14 sons of these 1946 imports, which included sons of *Rio Negro*, *Aimore*, *Arauto*, *Precioso* and *Gaucho*." Another ranch that used offspring from this importation was the 3X. Here Joe and Emily Dial have maintained three separate herds, an Indu-Brazil, a red Brahman and a Mucho Grande or grey Guzerat. Five of the first calves sired by bulls from the 1946 Importation were used later to establish their Indu-Brazil herd.[100]

Recent Brazilian Importations

While most American Brahman breeders appear to feel that there is already ample foundation stock in the United States to perpetuate excellent breeding programs, some breeders and speculators have continued to desire new animals from Brazil.[101]

As early as 1946, a group of Brahman breeders, headed by J. W. Sartwelle, went to Washington to lobby for the establishment of an international quarantine station for facilitating the importation of more cattle. Supported by the American National Livestock Association the ABBA and a number of other cattlemen associations, this group was able to receive the endorsement of the Bureau of Animal Industry. The necessary Congressional legislation was passed, and the bill, designating Swann Island in the Caribbean as the site for a facility, was signed by President Harry Truman. However, the bill was repealed by Congress in 1949.[102]

In was not until 1979, that a U.S. quarantine station was eventually built. It is located at Fleming Key in Florida, and is known as the Harry S. Truman Animal Import Center. The first shipment of 60 Brazilian *Bos indicus* cattle were released for distribution in July of 1980. Buyers from nine states were issued import permits and have since received these Brazilian cattle.

In April of 1981 eighty-one more *Bos indicus* cattle were imported from Brazil. Fifty-seven were Nellore.[103]

During the years, some *Bos indicus* cattle have also come into the United States from Mexico. Some of the most recent Mexican imports included 29 Indu-Brazil type cows and three Indu-Brazil bulls which entered South Texas in January, 1979, imported through the partnership of Garcia-Martinez near McAllen, Texas. Billy Thomas of Raymondville, Texas, who earlier formed a partnership with the famous Maitret family of Mexico, also imported several Indu-Brazil type cattle in the mid-1970's to Texas.[104]

While the importations of the 1920's, particularly that of 1924, and the subsequent importations have been very important in the development and evolution of the American Brahman, the significance of the Borden-O'Connor Importation of 1906 should not be forgotten. As J. D. Sartwelle recently pointed out:

> If it had not been for the good foundation stock established from those cattle imported directly from India, there would have been no stock with which to cross those Brazilian bulls. A good nucleus of Brahmans, conditioned by the coastal environment, already existed when those first 90 bulls were brought into the United States in 1924.
>
> Besides having helped produce notable foundation herds, the 1906 Importation also served to acquaint many coastal cattlemen with the adaptibility and prepotency of the Brahman breed.[105]

5

A Preference for Reds

As mentioned previously, some of the breeders of American Brahmans have preferred an animal that is red in color; consequently, approximately fifteen to twenty percent of the Brahmans registered by the ABBA are referred to as red Brahmans. It is believed that the Gir breed has been the chief transmitter of the red color, but the Guzerat and the Indu-Brazil have also been important in the development of this strain or type of American Brahman. The genesis, in fact, of the first red Brahmans in the United States is amorphous in character, but like all other American Brahmans, they were bred first of all for marketable qualities.[1]

While most red Brahman breeders have bred toward a near solid, deep red to brown colored animal, the speckled or even "pinto" colored type is popular among some breeders and buyers. The solid red, for instance, is preferred in parts of South Africa where the American Brahman has been a popular import item. On the other hand, some breeders along the American Gulf Coast like red and even black speckled cattle.[*][2]

*There are approximately 200 to 300 head of breeding age females of the black and black speckled strain in the United States, but there has always been a strain of black cattle in India. They are generally considered to be first cousins of the Gir. "Many Gir cattle in Brazil were mixed with the black and white Indian cattle, and occasionally one can see black and black speckled animals in fine Gir herds in Brazil." (See *Brahman History in the Makin'*, ABBA materials, Oct. 14, 1977.)

Practically all of the *Bos indicus* cattle that were imported into the United States during the last century and during the early part of this century were Guzerat and Nellore with some Krishna Valley breeding. However, some reports indicated that Gir blood was in a few of these first imports. Cattleman Milby Butler of League City, Texas recalled in 1945 that sometime between 1879 and 1883 his father bought a black bull that consistently sired brown or red calves. And Henry Clay Koontz of Placedo, Texas said that part of the basis for his red Brahman herd goes back to the Brahmans that his great-grandfather, Captain John N. Keeran, owned in 1878.[3] But certainly one of the most celebrated early red colored Brahmans was the Al M. McFaddin bull named *Prince*. This was the bull that McFaddin purchased at the 1904 World's Fair in St. Louis. A number of the earliest foundation red Brahman herds were descendants of *Prince*. *Sam Houston*, a red bull calf, was the first animal to be registered by the ABBA, and was believed to be descended from *Prince*.[4]

Sometime between 1913 and 1917, Milby Butler bought two of the McFaddin bulls. Although one was actually grey, both sired red and brown calves, and he used both in establishing a red Brahman herd. Later, Dr. W.S. Jacobs also purchased a herd of 50 red Brahman heifers, probably descended from *Prince* as they, too, came from McFaddin. Joe Hughes eventually acquired these same 50 red heifers, and he used them to develop his red Brahman foundation herd. Subsequently Hughes traded some 25 of these to Walter Fondren of Houston.

The Fondren Estate's red herd was one of the most important foundation herds in the development of the red Brahman. In 1947, Walter Fondren had purchased the Brazilian bull, *Arauto*, from Joe Pate. With *Arauto*, and red cattle purchased from Joe Hughes, which included two sons of *Estrello*, Fondren was able to develop a red American Brahman type which was admired by many breeders.

Fondren's great bull, *Arauto*, was later purchased by "Cy" Yoakum, but the remainder of the Fondren Estate's red herd was donated to the Methodist Hospital at Houston, Texas by Walter Fondren's widow. Before these cattle were transferred from the Fondren Ranch, the Methodist Church sold them to cattleman Ben H. Carpenter of Dallas, Texas and Senor Lecuna of Venezuela in 1959. Carpenter incorporated these red Brahmans into his herd at Dallas, but Lecuna used his reds to establish an important herd of red Brahmans at the Santa Elena Ranch of Madisonville, Texas.[5]

In a recent interview with Steve Allen, Santa Elena's capable manager, Allen explained:

> At Santa Elena we breed for beefiness more than anything else. We place little emphasis on features, such as the shape of the head, the ears and other similar things. Our ideal cow really is a red Brahman with the conformation normally desired by grey Brahman breeders. We like to see it thicker across the back and not as long-legged as some red Brahman breeders prefer.
>
> We like red Brahmans generally because they transfer a uniformity of color in crossbreeding programs. Ninety percent of the time, you can predict the offspring's color.
>
> It has been our experience in selling red Brahmans in South America that buyers there prefer the qualities normally identified with the American Brahman. Those we have dealt with, at least, are not interested in animals with predominantely Gir features.[6]

Another early Brahman breeder, who acquired descendants of *Prince*, was William C. Steinmann of Yoakam, Texas. Interestingly, Steinmann crossed six bulls sired by *Prince* with offspring of the old Frost-Barrow herd previously mentioned. However, it was 1923 before the Steinmann's actually established a red Brahman herd.[7]

At least one of the bulls imported by the Pierce Estate in 1906 was a red bull later named *Adam*. From this bull, the Pierce Estate established its first strain of red Brahman cattle, however, cattle from the 1924 Importation probably played a more important role in establishing their red Brahman herd. A grey bull that Milby Butler bought from the Pierce Estate in 1919 sired some calves that were light brown, and in 1920, Fred Locke sold a red spotted bull, probably descended from *Adam*, to William C. Steinmann. Stanley Kubela of Palacios, Texas also subsequently purchased a red bull descended from *Adam*.[8]

Bloodlines from these early importations continued to produce an interest in and a source for red colored Brahmans, but

A group of purebred red Brahmans enjoy sunlight on a South-central Texas ranch.

the 1924 and 1925 importations provided the first significant bloodlines for developing red Brahman herds. Like earlier importations, these animals were mostly of the Guzerat or Nellore breeds, however, a number of important Gir and Indu-Brazil types or crosses were a part of these groups. Some of these were *Martino* (a speckled bull), *Faceiro* (a medium sized bull of reddish brown color), *Hopkins* (a light red Guzerat bull with a touch of Gir breeding), *Dick* (an Indu-Brazil type bull and sire of *Red Dick*), *Estrella* (a Gir bull which was important in developing the Red Gir lines of many early herds), *Solid Red* and *Sombrero* (solid red bulls), *Calcutta* (a Gir bull), *Bam Rapaz* (alias *Fripo*, a Gir bull of red dappled color).[9]

Another early Brahman breeder, L.P. Bunge, founded his herd with a group of 50 heifers that he purchased from Peter Hahn. The cows were descendants of the aforementioned Hagenbeck Circus bull. Bunge developed his foundation herd by the use of *Martino* (a Gir bull) and *Faceiro* (an Indu- Brazil type bull imported in 1924). "Descendants of this herd were also used in the founding of the later Red herds to be established."[10]

As previously mentioned, in the Morias Importation there were also "18 full blooded Guzerat heifers, seven of which were already in calf to *Red Imes*, probably a full blooded red Gir bull. (See Chapter IV, "Those Brazilian Brahmans.") These were acquired by Dr. Jacobs who crossed the red seven with *Solid Red, Estrella* and other bulls from the 1925 Importation.[11] Jacobs also infused some red Brahman bulls from the Pierce Estate into this breeding. Later he sold this red Brahman herd to W.W. Moore of Bastrop, Texas who eventually crossed this bloodline with *Rio Negro* and *Rio Pardo* of the 1946 Importation. (See Chapter IV, "Those Brazilian Brahmans.")[12]

Red Dick, an offspring of the 1924 import, *Dick*, was bought by Milby Butler, "and for some reason," according to Butler, ". . about 50 percent of *Red Dick's* calves were silver grey or steel grey." But by using descendants of *Red Dick*, and red bulls from the Pierce Estate, Dr. T.M. Neal also established a red Brahman herd having 700 females, at one time the largest herd of red Brahmans in the United States.[13]

From the 1924 Importation, Henry Kimball of Cotulla, Texas purchased *Fripo* and *Conquista*. He also obtained some red cows from the Sartwelles and the McFaddins, later breeding *Fripo* to them. Kimball's herd, considered by some as one of the best red Brahman herds at the time, also contained blood from *Calcutta* and *Adam*. After the 1946 Importation, Kimball leased the bulls *Repucho* and *Rio Pardo* from the Garcias and bred their female offspring to *Gaucho* and *Precioso*. Daughters of *Rio Pardo* were bred to *Gaucho*, and *Indo-Gaucho* was one of the red herd bulls of this mating.[14]

Of the Brazilian importations that came via Mexico, none was more controversial than that of 1946. It produced a highly charged issue that the ABBA struggled with for a long time. The question was whether or not the 18 or so Brazilian bulls of the 1946 Importation should be registered with the ABBA. These bulls would play an important role in the development of red

A number of bulls from the 1946 Importation played an important part in developing the strain of Brahmans referred to as "red Brahmans." Here Dr. Nicholas E. Dutro, member of the Bureau of Animal Industry of the U.S., leads bulls of the 1946 Importation across the Joe Pate Bridge at Hidalgo, Texas from Mexico.

Brahman herds, but as mentioned earlier, the books of the ABBA were closed at the time for the registration of new foundation cattle. On top of this, the policy of the ABBA then was not to recognize Brazilian cattle unless they had been previously recorded in an official Brazilian association, While it was claimed that these bulls had been exported to Mexico (Sacrificio Island) under the supervision of the Triangular Society of Uberaba, Brazil, no papers could be produced at the time.[15]

Also, with the importation of these Brahman bulls, came renewed fears of aftosa outbreaks similar to those that followed the importations of 1924 and 1925. Several states, including Florida and Louisiana, threatened to place a quarantine on all Texas cattle if new importations were encouraged. Although these bulls had passed all quarantine restrictions set up by the Mexican and U.S. Governments, U.S. cattle interests outside of Texas often placed the blame for all infectious livestock diseases on Texas cattle. It was only after the Livestock Sanitary Commission of Texas requested that the owners of these new imports isolate their herds for a period of six months, that the fears of cattlemen in other states were abated.

The ABBA also changed its rules regarding imports and foundation cattle, and after the appropriate appraisals, these bulls of the 1946 Importation were finally registered.[16]

When an outbreak of hoof and mouth disease took place in 1947, the American-Mexican borders were closed for many years to the importation of cattle. At least one U.S. breeder, Fausto Yturria of Brownsville, Texas, was unable to bring in bulls he had presumedly purchased before the embargo went into effect. These bulls were *Baricuru*, *Realejo*, and *Dolar*. In order to meet this problem, Yturria sent his cows across the Rio Grande, had them bred to his bulls in Mexico and brought them back into the United States. From this operation, he was able to establish a Gir and Indu-Brazil type herd.[17]

"Descendants of the Kimball herd have been the basis of many of the red Brahman herds in the Cotulla area of Southwest Texas." Such breeders as Jess McNeil and J. Martin used Kimball Reds for their foundation stock. When Kimball died, the majority of his herd went to the Coquats of Encinal, Texas. The Holsworths of Collegeport, Texas acquired the red herd bull, *Indo-Gaucho*, as well "as a few famous daughters of *Gaucho* out of *Rio Pardo* cows."[18]

One of the oldest continuing herds of red Brahman cattle in the United States has been owned by the Clifford Brothers of Mercedes, Texas. The Clifford herd was actually established in the early 1940's with a foundation of cows from Joe D. Hughes, Milby Butler, the Bunge Estate and Henry Kimball. Alton and Mike Clifford bought their first red Brahman bulls from Milby Butler, Tom Hester and the Fondren Estate. According to the late Alton Clifford, "Red Brahmans were going down a dead end street at that time . . ." as there was no herd sire material available to upgrade herds. But after 1946, they secured the services of *Rio Negro*, *Rio Pardo* and *Gaucho*, all owned by the Garcias. "Of these, *Rio Negro* was generally considered to be the best . . . and today, his blood is prominent in many red Brahman pedigrees. Sometime after 1946, the Cliffords also obtained blood from two of the Guerra and Sons' bulls, *Gibimba* and *Canario*. Since 1970, they have concentrated on the bloodlines of *Rio Negro* and *Arauto*. Like many Brahman breeders, the Cliffords see that "their herd is calved year-round so that cattle of all ages are available for buyers from the climates and terrains of many different parts of the world."[19]

Another important Brahman breeder, sometimes referred to as the "Father of the premier Red Brahman," was the late C.E. "Cy" Yoakam. "A fascinating and ingenious man, Yoakam . . . made unparalleled progress in putting together and developing a line of red Brahman cattle during a short period of time."[20]

Starting in the Brahman business in the autumn of life, Yoakam always seemed to have his own idea of what an ideal Brahman should be, and he was able to accomplish a great deal in only a few years. In developing his foundation herd, he drew from a wide number of breeders, including the Pierce Estate, the Steinmann Brothers, Hogue Poole, the Clifford Brothers, Joe D. Hughes, Milby Butler, the Garcia Brothers, the Fondren Estate, and Henry Kimball. According to longtime Brahman breeder Henry Clay Koontz, "Yoakam paid unheard of prices . . .for the purchase of his foundation herd." One of the herds he purchased from the Garcias represented the breeding of *Repucho*, *Precioso*, *Capichaba*, and *Brilhante*.[21]

One of Yoakam's theories about raising red Brahmans was that it was important to keep about a quarter of Indu-Brazil strain in a red herd. "His idea was that the Gir would contribute the conformation, milking and fleshing; the Indu-Brazil would contribute the size and statue." It should be remembered though that the red Brahmans available to American Brahman breeders at this time were probably not "pure blooded Girs." Referred to as "Texas Girs" by Henry Clay Koontz, he explained:

> They were a Gir type having considerable Guzerat blood and were larger boned than the pure Gir that some breeders seem to want today. It was the Guzerat that gave the better Red Brahmans their good feet and desirable symmetry.

> Even those cattle during the early 1940's that were often referred to as Indu-Brazil were not pure. Most of these early 'Indus' were, in fact, bulls put on cows that were American Brahmans. It is even questionable about the mixture of the 1946 Brazilian bulls that were thought by some breeders to be pure Indu-Brazil. *Arauto*, for instance, was probably a seven-eighths graded up bull. It is only recently that people have started speaking in terms of pure Gir, pure Guzerat, etc.[22]

One of Yoakam's most famous herd sires was *Rio Red King*144, ABBA 46225, which was bred by the Clifford Brothers and sold as a calf to Yoakam. *Rio Red King* was out of *Madame Queen* and *Rio Negro* and part of the bloodline was from *Estrella.*[23]

Some of Yoakam's ideas on breeding cattle "shocked and astounded many," but he stuck by his theories, believing strongly in the future of the reds. "His greatest desire was to be a contributor to mankind by leaving a legacy of cattle which would be of value in producing meat and milk for the betterment of the generations to come."[24]

Part of Yoakam's early success was the fact that his spread, the Cherokee Ranch, was located on the famous Limestone Lift in San Saba County, Texas. In this area, natural soil nutrients have made it possible to produce near ideal pasturage for producing heavy bone structured cattle.[25]

Henry Clay Koontz was greatly influenced by Yoakam's techniques when he began concentrating on a red Brahman herd in 1960. At that time, Yoakam offered to lend Koontz one of his finest red Brahmans to use free for one year. Koontz remembered Yoakam saying, "I want to put the tops with the tops." The first calf crop made a "confirmed red breeder " out of Koontz.[26]

Of all the red blood lines that Koontz introduced into his herd, *Cherokee Vernon* 489 did the most to establish the type he was seeking to produce. Bred by Yoakam, *Cherokee Vernon* 489 was first loaned to the United States Sugar Corporation at Clewiston, Florida where he spent two years in breeding. On returning to Yoakam's Cherokee Ranch, Yoakam claimed that *Cherokee Vernon* 489 produced in one calf crop more herd sires than any other bull had done before or since. Koontz later bought this bull which stayed on his ranch until it died.[27]

Some of the so-called red Brahmans are actually speckled, such as this young bull on the Cherokee Ranch in South Central Texas.

In one year ". . . the progeny of *Cherokee Vernon* 489 won international championships in Guatemala, and El Salvador and won the coveted Gold Cup Award over all breeds at the Vryburg Agricultural Show in South Africa."[28]

Another famous descendant of a "Cy" Yoakam cow was *Miss 3X Sophisticated Lady* 5/5. Sired by *Mucho Grande*, she was the first red Brahman female to be named a national champion cow (1978) and she was never defeated in her class.[29]

"Cy" Yoakam's Cherokee herd was sold by his widow in 1969 to Tom Dean and C.L. Gray, both of whom had worked for Yoakam on the Cherokee Ranch. Dean and Gray leased the Cherokee, keeping Yoakam's herd intact for awhile; however, Gray eventually sold his interest to Dean. Dean in turn sold eighty percent of the Cherokee herd to Dr. Lee McKellar of Mount Pleasant, Texas to ease his financial indebtedness. This left Dean with only 52 cows and 18 heifers, predominantly the offspring of *Cherokee Arauto* 372, a son of *Arauto* and a Garcia Brothers Indu-Brazil heifer. "Most were grandmother-mother-daughter combinations . . . from this important nucleus the present Cherokee Ranch herd has descended."[30]

In his selection of females, Dean has stressed mothering ability, fertility, disposition, ability to reproduce themselves and conformation - all in that order. He recognizes the importance of conformation, but he doesn't believe that it is yet a dependable characteristic for reproducing itself. He feels that better conformation is sometimes linked with bad disposition and poor fertility.

Dean said:

> This is something we will have to work out. We can do it if we're honest in our culling and selection. Then in time, when conformation becomes a more dependable genetic factor, it can move up the ladder. It took ten years for us to eliminate bad disposition from these cattle. It took five years for us to bring bad bags under control, from fifty percent down to five percent of the females . . . you've got to sacrifice something when you're set on eliminating a bad fault.[31]

This hard line of culling has helped make the Cherokee herd a good source of purebred seedstock.[32]

Some of the best red Brahman blood in the United States was utilized by Stanley Kubela during the 1940's and 1950's. Kubela actually started with some Brahmans from the Pierce Estate soon after he finished a military hitch during World War I, but it was not until the early 1940's that he set out to establish a red Brahman herd. Kubela and his wife managed to acquire twelve

Red Brahmans have become popular in many parts of the Southeastern United States and in other parts of the world. These bred replacement females with a red bull are near Aliceville, Alabama (Picture courtesy of A.B. "Arnold" Lewis, Aliceville, Alabama).

cows from the Sartwelle Estate, all of which were reddish in color. "For years after, they occasionally were fortunate enough to select heifers of reddish color from the Sartwelle Brothers . . ." They also purchased animals that carried the blood of *Martino, Faceiro, Hopkins, Solid Red, Red Dick,* and *Rio Negro,* and for a time they leased *Rio Negro* from the Garcias. *Arauto* blood was also later introduced.[33]

One of the Texas red Brahman breeders that has relied heavily upon the blood of Dr. Jacobs' imported animals has been W.W. Moore of Bastrop, Texas. In time, Moore secured all of the red animals that had been developed by Jacobs, including several of the touted "AA" sires. One of his best red combinations has been Jacobs' blooded cows bred to *Rio Pardo.*[34]

In 1947, the M.S. Holsworth Ranch leased a son of this same *Rio Pardo* for one breeding season. And they have also utilized bloodlines from *Heroie, Precioso, Arauto, Rio Negro, Gaucho* and the Cherokee Ranch's *Cherokee Arauto* 442, son of *Arauto.*[35]

One of the largest registered red herds in the country today is owned by Dr. Lee McKellar at Mount Pleasant, Texas. McKellar established his breeding herd in 1960 with the purchase of 39 cows from W.W. Moore. Soon after, cattle from the Cherokee Ranch, the Clifford Brothers and the Garcias were added to this herd. "A landmark year for the McKellar Ranch was 1968 when the renowned sire *882* was purchased from the Cherokee Ranch." At the time, it brought $15,000. It is estimated that offspring from *Cherokee Negro* 882 on the Cherokee and McKellar Ranches are worth more than $1,000,000 today.[36]

A transplant program was started recently at the McKellar Ranch with the purpose of breeding more rapidly and selectively. According to records at McKellar's Transplant Center, the survival rate on transplants has increased from 25 percent to nearly 50 percent since this program was initiated. The McKellar Ranch has also helped the Hacienda Pachonte of Guatemala establish an embryo center.[37]

While red Brahmans are not nearly as widely used as grey Brahmans, they have still been particularly popular among breeders who like certain traits normally identified with the Gir and Indu-Brazil types, or who want to have a uniformity of color transferred during crossbreeding.

There is no unanimity however, among red Brahman breeders regarding the exact amount of Gir or Indu-Brazil blood desired. The three-quarter Gir and one-quarter Indu-Brazil formula once advocated by "Cy" Yoakam is popular among some breeders; however, it should again be remembered that pedigrees from Yoakam's Red Brahmans reveal the influence of what had become known by some as the "red American Brahman."[38] "Some red Brahman breeders, however, seem to be breeding more and more away from an earlier prototype and are attempting to emphasize features identified with Gir and Indu-Brazil breeding."[39]

Many red Brahman herds today are found in East Texas, but they can also be seen in Louisiana, Florida, Arkansas and other Southern states. Top quality red Brahmans have been a popular export item through the years, especially in Central and South America and South Africa. There has also been some demand for *reds* in Australia. (See Chapter IX, "A Breed for All Seasons.")[40]

6
Brahmans Pass All Tests

Early Research

When the first Indian cattle were imported to this country during the 1800's, there was at first no attempt made to establish purebred *Bos indicus* herds. These animals were used principally for crossbreeding purposes as American cattlemen were already aware of the importance and value of what is now called hybrid vigor; consequently, all of the early "Brahman research herds," or those herds for which records were kept on such things as growth rates, weaning ages, fertility levels, maternal traits, disease and heat tolerance, were crossbred herds with a high percentage of *Bos indicus* blood. "Most of the early Brahman bulls carried about three-fourths to seven-eighths of Brahman blood, and they were usually crossed with native stock or British breeds."[1]

Early research on Brahman cattle in America was not instituted by any public agency. As is so often true, it was begun by businessmen, in this case ranchers, who needed some fast, straight answers to some practical questions about profit and loss. Abel Pierce Borden was one of the first Brahman breeders to have enough Brahman cattle to make any kind of valid study. Immediately after bringing his imports from India to the Pierce Ranch in 1906, Borden had Dr. Mark Francis, Dean of Veterinary Medicine at Texas A & M College, manually innoculate the cattle with the Texas fever organism, determining then and finally the worthiness of the Indian cattle as a resistor to this dread disease which had proven so finally fatal to European cattle.[2]

Three years later at an annual meeting of the American Breeders Society at Omaha, Nebraska, Borden gave a rather detailed report on the tolerance of his new Brahman crossbreeds to insects, their faculty to gain weight on native pasture, and their rustling ability. Included in this report was the following:

> The object of the importation was to find a race of animals which would resist the insect pests of this section of the country, the principal one of which is the tick which causes Texas Fever. These bulls were bred to the best cows on the Pierce range and to a bunch of Hereford cows in 1907 and again in 1908. The crop of calves in 1908 were fine and did remarkably well . . . We would occasionally find some ticks on the bulls, but were never able to find a fully developed tick on them. The first crop of calves . . . is from 14 to 20 months old. They have been in tick infested pastures with ticky cattle all their lives, but they fail to carry any ticks to maturity, as far as I can see . . .
>
> This first crop of calves, about 300 in number, has grown up on the range as all our cattle, and they are fully 50% heavier than our ordinary range calves. They are as heavy as the calves a year older out of the same mothers but sired by purebred Hereford bulls. The crossbred animals are smooth with very strong constitutions, are good rustlers of rapid growth, and are animals that have courage enough to look you in the face when you go about them. The results are better than I expected.[3]

In a letter Borden wrote to Congressman R.F. Broussard in 1912, he continued to herald the virtues of his Brahman and Hereford crosses:

> The fourth crop of calves is now being dropped and the first crop of heifer calves are also bringing calves. We find that the crosses with the Hereford cattle are better than any other breed and have some specimans of such crosses that will attract attention even in the Hereford community.[4]

All of these animals are offspring from the Borden-O'Connor 1906 Importation. Borden said that he was never able ". . . to find a fully developed tick on them." He also said that when his first crop of calves matured, they were ". . . fully 50% heavier than our ordinary range cattle."

As mentioned earlier, part of the Indian cattle that were imported into this country in 1906 went to the estate of Thomas M. O'Connor. Some studies were also made on this segment of the herd by an entomologist, J.D. Mitchell, who was employed by the U.S.D.A.'s Bureau of Entomology. Mitchell was particularly interested in knowing what sort of resistance these cattle had to insects in Texas, especially the Texas fever tick. Having made studies of the fever tick some years before the importation, he was fascinated with the ability of O'Connor's Brahman cattle to remain almost completely free of ticks. His theory at the time was that the waxy secretion (though not greasy or offensive in

When Texas A & M began its early studies with Brahman cattle, it was difficult to find a purebred bull; so they started with a three-quarter bull like this one from the Pierce Estate. It was crossed with high grade Hereford cows.

odor) emanating from the hide of the Brahmans was offensive to the ticks and a number of other insects. He further recorded ". . . that ticky pastures used for Brahman cattle became more or less free of ticks after two or three grazing seasons."[5]

One of the first institutional research studies involving Brahmans was carried out at the Senora Ranch Experiment Station of Texas A & M College, starting in 1920 and running through 1929. Sponsored in part by the U.S.D.A.'s Bureau of Animal Industry, this crossbreeding program eventually involved three sites. They were Texas Substation 7 and 14, and the King Ranch at Kingsville, Texas. Initially heading these studies was Jay L. Lush of Texas A & M who was actively aided by J.M. Jones, R.E. Dickson, W.H. Dameron, E.M. Peters, E.W. Thomas, O.L. Carpenter, J.H. Jones and W.E. Flint.[6]

The first of these investigations involved the comparison of breeding, feeding and slaughter tests between the progeny of a high grade Brahman bull and a herd of high grade Hereford cows with the progeny of a purebred Hereford bull and another herd of high grade Hereford cows. Purebred Brahmans were almost impossible to find at this time; so a three-quarter blooded Brahman bull from the Pierce Estate was first used; however, this bull was later replaced with a Brahman bull borrowed from J.A. McFaddin that was of a 61/64 grade. The total study involved the breeding, birthing, feeding, slaughtering and selling of 444 calves.

In feed lot demonstrations at Station 14 beginning in 1921, records revealed that the Brahman crossbred steers and calves were always heavier than the Hereford steers and calves when they came to the lot. And after being fed out, they also had less shrinkage when moved to the market. These crosses also made slightly higher dressing percentages than the straight Herefords, running generally 2½% to 3% better.

Early tests also revealed that the Brahman cross hindquarters showed the highest percentage of the whole carcass weight. During these tests the bulls were very leggy and showed little resemblance to the beefier type American Brahman that most Brahman breeders prefer today.

Gross cash returns from steers born in 1921 and full fed for 120 days beginning November 1, 1922, actually showed a 23¢ per head advantage for the Hereford steers over the Brahman crosses. But one year later, the advantage was reversed with the Brahman crosses bringing on the average of $6.13 to $7.69 per head more, depending on their age. This trend remained with four separate calf crops tested after 1921, with the advantage of Brahman and Hereford crosses running as high as $8.83 on the average over straight Herefords in 1928. In that year, the average price paid for Brahman crossed steers from these experiment station herds was $98.86 per head as compared to $88.03 per head for Herefords. Interestingly, the prices for all cattle being sold from the experiment stations had approximately tripled since 1921, reflecting what at the time was a spiraling national economy.

Fertility Levels and Maternal Instincts

Observations were also made on the fertility levels of the Brahman bulls used during these same studies. The data collected at this time, "concerned only the relative success of Brahman bulls and Hereford bulls in siring two separate groups of cows which were kept under like conditions." According to Dr. J.L. Lush, "The observed differences were too small and too irregular to indicate any real difference in the percentage of calf crop sired by the different kinds of bulls."[7]

Although the fertility level of the Brahman bull has been questioned at times, there is certainly no consensus of opinion that his fertility level is abnormal. Like all breeds, "the general fitness to the environment directly affects the effective fertility level of the bull."[8] Thus the proven ability of the Brahman bull to walk and keep up with the herd, his ability to resist diseases, such as anaplasmosis and paraplasmosis, his virtual freedom from pinkeye, lumpy jaw and cancer eye, his exceptional endurance and certainly his unexcelled longevity all help to improve his lifetime fertility performance.[9] It has been claimed that longevity is not so important today as open cows are now easily replaced with fertile heifers. While this may be true in some cases, certainly longevity in prepotent sires will remain important.[10]

"Investigations by Patrick and others at L.S.U. showed that the fertility level of European dairy bulls is greatly reduced when the air shade temperatures go over 80°F. This is an important factor regarding fertility levels when we consider that most range

In studies run at L.S.U. and the Texas Agricultural Experiment Station, it was demonstrated that the fertility rates of Brahman bulls compared favorably with other breeds. As one leading Brahman breeder said, "A good performing bull is like an athlete. He has to be properly managed and always in good shape regardless of his breed."

land areas of the world have average daytime summer temperatures that are 80°F or over. Indeed, in a number of the areas of the world where the American Brahman has been introduced, summer daytime temperatures are often over 100°F, and cattle without some Brahman blood simply cannot survive.[11] (See the Austrialian studies by R.B. Kelly and the Tingo Maria Studies in this chapter.)

Commenting on male fertility rates, long time Brahman breeder Edgar Hudgins said, "A good performing bull is like an athlete. He has to be properly managed and always in good shape, and largeness is no criteria for judging fertility. A medium sized American Brahman bull that is in good condition is as fertile as any other bull."[12]

In 1954, a joint project carried out by the Texas Prison System, the Texas Agricultural Experiment Station, and the Brazoria County (Texas) Cattleman's Association tested the fertility levels of Brahman, Angus, Hereford and Santa Gertrudis bulls when put on graded Brahman cows. Twenty half, or more than half, Brahman cows were assigned to each bull. At the end of the year, the Brahman bull had sired 16 calves for an 80% calf crop. These calves averaged 465 pounds for a total production of 7,435 pounds for the cows bred to the Brahman bull. On the other hand, 7 calves were credited to the Angus bred cows for a 35% calf crop and a average of 460 pounds per calf at weaning with a total production of 3,220 pounds. Twelve calves were credited to the Hereford bred cows with a 60% calf crop that weaned at 445 pounds and had a total production of 4,895 pounds. The Santa Gertrudis bull was the nearest in total production to the Brahman sire. This bull sired 14 calves for a 70% calf crop that averaged 445 pounds at weaning, making a total production of 5,785 pounds.[13]

In a similar study started in 1963 by the University of Florida's Experiment Station at Brooksville, Florida, highly graded unregistered Brahman cows were shown to have an unusually high fertility rate when bred with American Brahman bulls. According to Dr. Marvin Kroger, "These graded Brahman cows have demonstrated an extraordinarily high fertility level, higher than any cows, including the Angus, that we have investigated at the Brooksville Station. There overall calf production has consistently been over 80% in a fifteen-year-period."[14]

In a project funded principally by the American Brahman Breeders Association and the Texas Agricultural Station at Overton, Texas between 1975 and 1979, rather intensive studies were completed on the estrus cycles of Brahman, Brahman-Hereford crosses and Hereford cows. These results showed that the Brahman cow often needs more estrogen to come into heat, has a lower output of the hormone causing ovulation, and ovulates earlier in relation to "standing heat," thus causing lowered conception rates. However, it is suggested that the selection of Brahman cows for a larger corpus luteum (the tissue in the ovary that secretes a necessary hormone) levels and higher progesterone levels could increase conception rates equal to those found in other breeds.[15] (See Cartwright, Fitzhugh studies on fertility levels in this chapter.)

One of the factors that may have an inhibiting factor on the estrus cycle of the Brahman cow is their generally longer lactation period. However, many Brahman breeders will tell you that this is a mixed blessing as this extended lactation period helps explain why Brahman crossbreeds are so famous for their excellent maternal qualities.[16]

Certainly one of the most important maternal attributes of the Brahman and Brahman crosses has been the ease by which they calve. "They are practically devoid of calving problems associated with the size of the calf. Except in rare cases of malpresentation, assistance in calving is not necessary, and it is quite common for purebred breeders to calve two hundred cows with no problem in this respect."[17]

Generally the Brahman-type calves are small at birth, seldom weighing more than 70 pounds. "Along with a fantastic mothering ability, this is one of the important characteristics hybrid cows seem to inherit from their Brahman parent."

Records kept at Manyberries Station in Canada showed that none of the half-Brahman crossbred heifers required assistance at calving, either as two year olds or at any subsequent calving. As Leon Locke pointed out some years ago, "In areas of the world where adequate skilled help is not readily available, the ease with which Brahman and Brahman-type cows calve is a great inducement to their use."[18]

Walter Hudgins once said, ". . .if a man wants to put weight on his calves he needs to get them out of Brahman cows. They're the finest mothers in the world. They'll calve, and be up in an hour and take the calf out in the sun. We'll get an 80% calf crop, compared to maybe 50% that some of my neighbors with

A twenty-one year old F_1 Brahman-Hereford cow with her eighteenth calf at the East Texas Pasture Laboratory, Lufkin. The lifetime average adjusted 205 day weaning weight of her calves was 505 pounds.

Northern breeds report. And there'll be plenty of milk for the calf right through our hot weather that drags down other cows."[19] It has been said that "the ideal lactation of a beef cow would be the ability on fair range to produce an amount of milk of the exact fat content daily to maintain the calf in the most rapid rate of growth and state of good health for the entire suckling period of its life, leaving the calf in such a degree of fatness that it can either be slaughtered or that it can go through the adjustment of weaning without suffering severely, and at no time producing more milk than the calf is able to consume."[20]

Most studies have shown that American Brahmans and Brahman crosses produce more milk than do cows of the three major European beef breeds. Importantly, "this milk flow is sustained by the Brahman cows over a longer period of time." "Spoiled udders" that develop when more milk is produced at the time of birth than is necessary is not normally a problem with the American Brahman or Brahman cross. "Milking out," a distasteful job to the cowboy, is hardly ever a problem with Brahman cattle.[21]

When temperatures rise or soar during the spring or summer months, the Brahman cow generally maintains a constant flow of milk. This, of course, is also significant as it does not force the calf to forage for itself; on the other hand, among European breeds milk flow often tends to drop as temperatures rise. (See section in this chapter entitled "Brahman Blood Boosts Dairy Breeds.")

Purebred Brahman cows have not been compared with purebred European or English cows from a calf producing standpoint on a large scale. However, studies consistently show that calves out of Brahman-European and Brahman-English crosses weigh considerably more at six months of age than calves from cows without Brahman blood where the conditions are similar.[22] (See Chapter VII, "Brahman, the Crossbreeding's Common Denominator.")

One of the factors sometimes overlooked in considering reproduction disadvantages that may be attributed to the Brahman are environmental conditions. In other words, "Unfavorable, limited breeding seasons plus seasonal and other calf mortality problems have at times been contributing factors . . . by far the major culprit has been the harsh environment in which the Brahman cows must usually have to produce. Hot humid weather, heavy infestations of internal and external parasites, and . . . low quality forage characteristic of the Gulf Coastal United States are adverse to high production levels in growth and reproduction."[23]

Dr. Tom Cartwright and Dr. Henry Fitzhugh of Texas A & M conducted a series of studies during the early 1970's in which reproductive records of different purebred Brahman herds with different environments were compared. The herds involved where those of the W. H. Stuart Ranch at Bartow, Florida, the J. T. White Ranch at Hearne, Texas and the ranch of I. W.

Dr. Tom C. Cartwright

Dr. Henry A. Fitzhugh, Jr.

Whitaker at Donie, Texas. All herds were large enough to provide numbers for meaningful results, but small enough so that cows could be observed on a daily basis. While the nutritional needs of all cows were met, the general care was no greater than would be expected on any well-managed operation; however, the White and Whitaker herds, located just outside the Gulf Coast area, were somewhat more intensively managed.*

Results from these studies actually demonstrated a very high fertility potential by a large percentage of the cows in all herds. However, there was some difference. In the more intensively managed herds, 75 to 80 percent of the cows had a calving interval of less than 365 days after the first calf, and many continued to produce calves on less than a 365-day interval in successive years. Those cows on the less intensively managed herds, as expected, had a somewhat longer calving interval.

Calf survival percentages were 90% or above for all three herds, with one of the more intensively managed herds running 98.1%.[24]

Cartwright and Fitzhugh concluded from this study and other investigative work:

> It is well established that the Brahman cow is slow to cycle when under nutritional stress, especially if she is under the additional stress of nursing a calf [but] the valuable survival mechanism that prevents her from cycling during stress also operates promptly to initiate cycling when nutrients are plentiful.[25]

With proper management it has thus been demonstrated that there need not be any real fertility problem with Brahman cows.

Useful longevity, "the age a cow reaches before becoming subfertile or infirm," is another factor that should not be

*One of the most fascinating Brahman operations in Texas is that run by J. T. White, former president of ABBA, and his wife, Katie White. The White's program is a relatively small one usually requiring less than 200 head of registered Brahmans which they run on less than a section of land, much of which they cleared by hand in the 1920's and 1930's. Mrs. White is an excellent bookkeeper, keeping track of every penny that comes in and goes out. Most of their profits have gone toward scholarship funds for needy students.

Brahman breeders J. T. and Katie White of Hearne, Texas with one of their "pets." Their Brahmans have been some of the most gentle cattle in Texas. J. T. White is a fourth generation cattleman, and he is convinced that the American Brahman is the best breed around.

overlooked when assessing the fertility or the number of calves the average cow in a herd raises to weaning in a year's time. But what is often overlooked is the number of calves the average cow in a herd raises to weaning age in a lifetime. Most studies thus far indicate that the "useful longevity" for Brahman cattle exceeds that of all European or British breeds. In fact, "based on

Mr. Anthony Doherty brought purebred Brahman bulls from the Borden Estate in 1915 to the Baton Rouge area. In 1928 he donated four heifers to Louisiana State University to start a crossbreeding program. Doherty insisted that Brahmans were essential to any successful cattle operation in Louisiana.

breed average, Brahman and Brahman-cross bred cows live and produce about 50% longer than European breeds."[26]

With proper nutrition and good management and with the advantage of natural longevity, "the Brahman cow can equal and in many cases exceed, the widely recognized levels in the British breeds."[27]

Other Early Brahman Studies Are Encouraging

At about the same time that the aforementioned Lush directed studies were phasing out in 1928, Louisiana State College, now Louisiana State University, was acquiring its first Brahman cattle. Harry Gayden, at one time in charge of the LSU beef cattle herds and later capable Executive Secretary of the ABBA,* recalls that ". . .in 1928 Anthony Doherty of East Baton Rouge Parish gave the University two unregistered Brahman heifers and encouraged officials to begin a crossbreeding program with them." Doherty's donated Brahmans were presumedly descended from some of the Pierce Estate Brahmans which he had been buying since 1915, again showing the importance of the 1906 Importation made by Borden and O'Connor.

Shortly after acquiring the two heifers, Professor E.L. Jordan, head of the LSU Department of Animal Industry, purchased a 15/16 Brahman bull from a Eugene McQuaid of Baton Rouge. "This bull was described as an 'excellent sire and quite a pet'," and was eventually used in crossbreeding with Hereford, Shorthorn and Angus cattle that were owned by LSU. Harry Gayden remembers with fondness that the McQuaid bull was extremely gentle and had a strong penchant for sweet potatoes.[28]

Dr. C.I. Bray, who was placed in charge of Brahman crossbreeding investigations shortly after, wrote, "From 1929 until 1933, records were kept on calves' birthweights and approximate monthly weights of the herd during the summer until weaning time." Unlike the earlier Texas Experiment Station studies, no difference in birth weights was first observed; however, a "considerable difference in growth in favor of the Brahmans. . ." did occur. Calves sired by Brahman bulls on Hereford and Angus cows averaged over 56 pounds heavier at weaning time than did calves by purebred Hereford and Angus sires. Yearling Brahman-cross heifers weighed about 84 pounds heavier, and the two-year-old Brahman crosses weighed 91 pounds heavier. "Occasionally the half Brahman calves weighed over 600 pounds at weaning, but the average ran slightly over 500 pounds as compared to a 436-pound-average for the other calves. Harry Gayden recently said, "What really sold me on Brahman cattle back then was the comparison of Brahman crosses with commercial Shorthorns and Herefords. One of the

*Mr. Harry Gayden ably served the ABBA during some of its most trying years, often acting as a conciliatory force in his own quiet way.

Harry Gayden, at right, was involved in early experimental work with Brahmans at L.S.U. Later he became Executive Secretary for the ABBA. Here he is with sculpturer Jim Reno who is working on a model for the Association in 1954.

progeny of the McQuaid bull, presumedly a ½ Brahman calf, weighed 602 pounds by the time it was 7½ months old. It was the largest calf at that age I had ever seen."[29]

One of the most satisfying discoveries to come out of these early LSU investigations, as far as Brahman breeders were concerned, was the results of the creep feeding experiments. In these experiments, which ran from May 5, 1933 to September 15, 1933, non-Brahman creep fed calves were compared to non-Brahman calves fed on pasture alone and Brahman crossbreds fed on pasture alone. As expected, there was a marked difference in weight gain between non-Brahman creep fed calves and non-Brahman calves fed on pasture alone. The difference averaged one-half pound a day per head. However, the

Rasputin Imperator **was one of the early Brahman bulls used at L.S.U. for experimental studies. This bull was 13 years old at time of picture, 9/15/50.**

Brahman crossbred calves that were left on pasture gained almost the same amount of weight each day, 1.77 pounds, as the creep fed Brahman calves which gained 1.78 pounds each day.[30] Because of the times and the added expense of creep feeding, these findings were obviously important to Brahman cattle owners at the time as they are even so today. But these experiments, perhaps, will have an even greater significance in the future as the world demand for grain foods increases and as grain becomes an ever more important source of fuel. (See Chapter IX, "A Breed For All Seasons.")

A report made in 1975 by R.L. Moore, H.W. Essig and L.J. Smithson, which appeared in the *Journal of Animal Science*, also indicated that Brahman bull calves were more efficient on low energy (high roughage) rations when compared to

Herefords. On the other hand, Herefords were shown to be superior on a high energy (low roughage) ration.[31]

In 1932, the research at LSU included a study of Brahman-crosses that were finished with grain. Two lots of mixed yearlings, containing four Brahman crosses, were fed grain on grass. When the records of the Brahman-crosses were compared "with four of equal age by beef type sires, the Brahmans gained 1.95 pounds per day as compared to 1.46 pounds for the others." These Brahmans also later sold at $6.28 on the New Orleans market while the others averaged $5.28.[32] This, of course, was 1932 when the prices for cattle were "bottoming out." The cash differences might be even greater today.

"In the early days," according to J.W. Sartwelle, "a man might have a Brahman bull, but you'd look a long while before you'd find any calves with preponderantly Brahman blood. The buyers always took them, because they were the fattest calves."[33]

After 1933, LSU suspended experimental work on this project for awhile, leaving their Brahman sire in the grade herd for several years to breed to purebred and grade Shorthorns. And for many years, many of the Brahman grades in the commericial herd at LSU were descended from these Brahman-Shorthorn and Brahman-Hereford crosses.

"Uncle Brad" Fairchild was one of Louisiana's early cattlemen who started using Brahman stock in the Baton Rouge area around 1880. Although his Brahman cattle were probably not used in the experimental work at L.S.U., the large herd of Brahman-crosses he ran along the Amite River stimulated interest in the Brahman. "Uncle Brad" was the grandfather of Alfonse Forbes of Greenwell Springs.

During 1933 Harry Gayden was appointed head of the LSU beef cattle herd. It was also in 1933 that the University secured a loan from Dr. W.S. Jacobs of several registered bulls through the efforts of Brahman breeders Brogan Fairchild and F.J. Whitehead of Port Allen. In the years that followed, LSU also received a number of registerd bulls from the J.D. Hudgins Ranch.

The year 1937 marked a milestone for the American Brahman industry as well as for LSU. That year the Department of Animal Science designated Harry Gayden to purchase a purebred Brahman herd for the University. Gayden travelled to Orange, Texas and picked up 20 cows from the Stark Estate.[34]

One of the most extensive early Brahman studies was made at the King Ranch in 1930. During that year, 6,591 Brahman-cross steers and heifers and 5,664 Hereford steers and heifers were grass fattened under virtually the same conditions. By the time that 1,739 of the Brahman-cross cattle had been fattened and readied for market, only 575 head of the Herefords were sufficiently fit for sale. By the end of the year, all of the Brahman-cross cattle had been sold or judged ready for sale while it was the end of the second year before all of the Herefords were sufficiently fat to market.

Over the whole consignment, the Brahman-cross steers averaged approximately $1.17 per head more than the purebred Herefords, and the Brahman-cross heifers averaged approximately $1.70 more than the corresponding Hereford heifers.[35]

Dr. C. I. Bray

In another similar study made at LSU under the direction of Dr. C.I. Bray and the help of Gayden, feeding tests using corn and dehydrated sweet potatoes were fed to Brahman-Angus crosses and non-Brahman cattle. The Brahman-Angus crossbreds gained on the average of 2.19 pounds per day in a feed lot as compared to 1.72 pounds for other calves not having Brahman blood. And in still another LSU study mentioned earlier, non-Brahman calves did almost the same when creep fed as Brahman-crosses did on range grass, with the non-Brahman cattle averaging only .01 more pounds per day when creep fed. When non-Brahman calves were put on grass alone, they gained only 1.27 pounds per day. Studies at the Georgia Coastal Experiment Station at Tifton, Georgia in the 1950's showed that Brahman X Hereford calves were 18% heavier at weaning than calves from Angus X Hereford calves.[36]

Although purebred Brahmans have sometimes been characterized as slow growing feed lot cattle, once they are crossed with British breeds, the progeny excell either the British breeds or the purebred Brahmans. When Brahman and Shorthorn purebreds and crosses of varying degrees were placed on three types of range that included native partly improved and improved pasture, at Ona, Florida in 1950, the Brahman-crosses performed better on all pastures. This performance included such factors as weight gains, fertility levels, calf weights, and rate of weight gain. Of the crosses, F_1 Brahman-crosses and back crosses (in both directions) demonstrated the greatest heterotic changes.[37]

While these studies were carried out in the warm coastal regions of the United States, other studies with Brahman-crossbred herds have been carried on as far north as Canada. Studies in Alberta ". . . reported by Peters and Slen in 1967 demonstrated a marked superiority in the performance of Brahman-crossbred steers and dams over what was considered to be the better performing British cattle." Also at the Canadian Range Experimental Farm at Manyberries, Alberta, "part Brahman cattle were tested to determine whether they would be adaptable to the . . . more northerly ranges of the country . . . half-bloods compared favorably with the European breed

Some of the earliest studies involving Brahman cattle took place on the King Ranch, and through the years, it has been one of the principal buyers of purebred Brahman cattle. This ranch was also the birthplace of the Santa Gertrudis breed.

observed in the same test and were superior in four of five categories studied.[38]

During the late 1920's and the early 1930's, the King Ranch became one of the principal buyers of purebred Brahmans. The J.D. Hudgins Ranch, for instance, sold over 600 head of Brahmans to the King Ranch during this period of time. Needless to say, the King Ranch's desire to buy purebred American Brahmans at this time was a great balm to an otherwise sick cattle market. These were the worst years of the Great Depression and the enthusiam for promoting the more expensive purebred cattle had waned. Not only was the King Ranch buying purebred Brahmans, but they were also shipping some to the East. Some of the Hudgins' Brahmans in fact went to the South Carolina ranch of Winston Guest, nephew of Winston Churchhill.[39]

Australian Studies

Another early and significant study on crossbreeds derived from American Brahmans was actually made some 9,000 miles from the North American Continent. This was the research that was begun by Dr. R.B. Kelly of Australia in 1933 which helped to establish the American Brahman as an important breed in the Land Down Under.

Bos indicus were not entirely new to Australia in 1933. In fact, numbers of Indian cattle had been imported to Australia from India during the early days of colonization. Similar to the earliest importations of Indian cattle to the United States, these cattle either died out completely or were absorbed by native breeds that were then more popular. Later imports made to the Northern Territory of Australia in 1843 and to the Essington and Darwin Districts in 1872 became the basis of a few crossbred herds. A large number of descendants from these cattle can be found today in the Darwin area of Northern Australia.[40]

But the first big attempt to import American Brahmans to Australia came in 1933 when a syndicate was formed made up of the C.S.I.R.O.,* and Messrs. Winter-Irving and Allison of Wealwandangie, the Queenslands Stations of Glenprarie, Mr. Wright of Waverly and the Meredith Menzies and Co. Ply. Ltd. of Millungera. These individuals and organizations decided to use part of their resources in bringing approximately 20 American Brahmans to Australia. During that year approximately 20 head of American Brahmans were purchased from several Texas ranches which included the J.D. Hudgins Ranch, the Dr. W.S. Jacobs Ranch, and the Pierce Estate. Thirteen of these Brahmans came from the J.D. Hudgins Ranch of Hungerford, Texas. One of these imported bulls was a Santa Gertrudis, but all of the rest were actually registered by the American Brahman Breeders Association. The names and original owners were:

*The C.S.I.R.O. was Australia's Council for Scientific and Industrial Research. A member of this organization actually went to Texas to purchase these cattle in 1933. (See *Brahmans' Australian History*, Australian Brahman Breeders Assoc., Rockhampton, Qld., 1971)

Supimpa, **a Hudgins' bull from the Ruffier Importation, was the sire of a number of the American Brahmans that were exported to Australia in 1933.**

Jake, ABBA 651 and *Ponieta*, ABBA 2094 from the W.S. Jacobs Ranch; *Red Quo Vadis*, ABBA 646, *Christmas Vulgas*, ABBA 648, *Ricardo Martin*, ABBA 647, *Himalaya* 1st 888, ABBA 643, *Supimpa* 1st 396, ABBA 644, *Commander Moroto, Jr.* 506, ABBA 645, *Dot Victor*, ABBA 2091, *Donna Jeremias*, ABBA 2087, *Ann Victor*, ABBA 2092, *Naomi Ideal*, ABBA 2090, *Rosa Amorin*, ABBA 2086, *Rena Martin*, ABBA 2089, and *Lillie Victor*, ABBA 2088, all of which were sold to the Australians by the J.D. Hudgins Ranch; *Hutchins*, ABBA 650, and *Louise*, ABBA 2093, both of which were sold by the Pierce Estate and *Abel*, ABBA 649, which was bought from A.P. Borden. None of the above cattle carried any of the *Manso* line, which was to play such an important role in the evolution of the Australian Brahman herd, as this was before his time.[41]

American Brahmans caught on fast in Australia, and many early imports were utilized in experimental work. This peaceful herd was photographed in Queensland, Australia, "the nation's beef state."

In time these 1933 imports of American Brahmans and their first progeny were crossed with Herefords, Polled Herefords, Shorthorns and Polled Shorthorns. The results in some areas was nothing short of miraculous as "cattlemen from all recognized production areas in Australia gave the Brahman definite approval . . . both as straight breeds and in many defined crossbreeding programs."[42]

Up until 1941, when Australia was threatened by an invasion from the Japanese, close studies were made of these imported cattle and their progeny. Like many early American coastal cattlemen, the Aussies buying these Brahmans were very much interested in knowing how resistant they were to ticks and other insects.

In only a few years this small importation had grown from 19 animals to 15,000 Brahmans and Brahman crossbreds. In time, Brahmans would spread to many parts of Australia, but initially most of this early expansion took place in areas where there was heavy tick infestation. In some of these regions, no cattle at all had been able to survive before the arrival of the American Brahman.[43]

The Brahman's resistance to ticks was dramatically shown to Harry Gayden some years ago when he was traveling by Landrover over the arid outbacks of Western Australia. Gayden recalled:

> We were hundreds of miles from a town, when we came upon a pond surrounded by a large herd of milling cattle. Most of them were part Brahman and were virtually free of ticks, but one that we noticed obviously carried no Brahman blood, and every part of its body appeared to be covered with ticks.[44]

The Tick's Aversion to Brahmans

Experiments conducted under Dr. Kelly's supervision showed that the American Brahmans and their first progeny were "for all practical purposes 100% resistant to Australian cattle ticks." It was Dr. Kelly's conclusons that "there was a direct relation between the degree of tick resistance possessed by the cattle and the percentage of Brahman blood they possessed . . . and that among the half-breds and quarter-breds the cattle with long hair coats were infested more severely by ticks than were those with short hair coats."[45]

Many studies have been made on the Brahman's resistance to ticks and other insects since Kelly's research. And all have shown that purebred and crossbred Brahmans have a remarkable resistance to ticks and other such insects as horn flys and mosquitoes. A more recent 15-month study by J.E. Frisch and J.E. Vercoe completed in 1978 reported ". . . Brahman as resistant to ticks, worms, pinkeye heat (and) nutritional fluctuations."[46]

In 1969, Phillip E. Tugwell, E.C. Burns and J.W. Turner of LSU reported a study on levels of Brahman breeding and color as factors affecting horn fly (*Haemtobia irritans)* repellancy. And it was found that as the percentage of Brahman blood was increased among cattle, "actual fly counts decreased, regardless of color." An Arkansas study with Brahmans, however, completed in 1977, indicated that Brahmans and Brahman-crosses with lighter colored coats resulted in a lower horn fly count - at least for that region.[47]

Another insect, which is not always considered as effecting the general performance of cattle, is the tiny but mighty mosquito. In 1973 and 1976, C.D. Steelman, T.W. White and P.E. Schilling completed studies on the efficacy of Brahman characteristics in reducing weight loss due to mosquito attacks. In this work Brahman steers were found more tolerant to mosquito attack than Herefords based on weight gains.[48]

The reason early coastal cattlemen were so concerned about the Brahman's resistance to ticks, of course, was due to the Spanish fever tick. But besides the fact that Brahman and Brahman-crosses carried few ticks, even those Brahmans bitten by the fever tick showed only a small reaction. Dr. H. Schmidt of the Veterinary Division of Texas A & M explained, "They will take tick or splenetic fever although in these tests it was in such a light form that a caretaker would hardly notice that they were sick at all."[49] Early studies made at the Pierce Estate revealed similar findings.

Cattle That Rev Down

One of the intriguing questions that many Brahman cattlemen have wondered about for a number of years has been why *Bos indicus* cattle and *Bos indicus* crossbreds can withstand heat extremes so much better than other breeds. Anyone who worked, raised or experimented with Brahmans was aware of their marked tolerance to heat and direct sunlight, but no one had satisfactorily explained why. Dr. Kelly of Australia set about to find some answers. One of Kelly's theories, that is still commonly held by many cattlemen, was that there was a definite relationship between the Brahman's ability "to sweat" and its heat tolerance.[50]

Little is known of the sweating ability of various types of cattle, but Kelly microscopically examined sections of some Brahman

Long noted for their great tolerance to heat, this picture graphically shows how other breeds will seek shade while Brahmans continue to graze in the hot sun.

skin. "In a microscopic field of .8 square millimeters he found an average of 9.33 'sweat glands' in the skins of one-half Zebu cattle, and an average of 5.25 glands in the same area of skin of one-fourth Zebu cattle. In the skin of the Holstein-Friesian, there were only a few glands and those were hard to find." Although this one study certainly fell short of offering anything conclusive about the heat tolerance of Brahmans or Brahman-crosses, other similar studies have been made since Kelly's study, and as Dr. Ralph W. Phillips, one-time chairman of the Animal Industry Branch of the United Nations, said in 1948, "The relationship between an animal's ability to sweat and its heat tolerance is probably more than mere conjecture."[51]

The question of whether or not Brahman or Brahman-crosses really sweat at all was recently raised by Dr. Marvin Koger of the

University of Florida. "Although Brahmans do not sweat in the sense that horses do, they do secrete or exidate fluids and there is a heat exchange," said Dr. Bill Turner of Louisiana State University.[52] But Koger contends that the great heat tolerance that Brahmans and Brahman-crossbreds have demonstrated is strictly a matter of metabolism, and that the Brahman has the ability to "rev down metabolic processes during extremes of heat." Early tests made at the Iberia Livestock Farms at Jeanerette, Louisiana on Brahmans, Brahman-crosses, and Angus cattle showed that respiratory rates, as well as body temperatures, went downward as Brahman blood was increased.

As far as "Brahman sweat goes," Dr. Koger explained recently, "the secretion that is so conspicuous on Brahmans during hot days is not perspiration at all. It contains few salts but it is rather a sebum or fatty liquid secreted by the sebaceous glands." Koger said that there are some rudimentary sweat glands that ooze out a little perspiration, but these glands are not so sensitive to heat as regular sweat glands are on a horse, for instance. And certainly they are not an important factor in explaining the Brahman's heat tolerance."[53].

One of the earliest studies made on *Bos indicus* hides was made by W.D. Hunter of the U.S. Bureau of Entomology in 1911. Hunter later reported the following:

> The Zebu hide is quite different from that of our common cattle. We examined hides from slaughtered animals and made scores of careful measurements of different parts of the hide. The Zebu skin is appreciably thicker than that of the Shorthorn. And the texture is different. There is an outer layer that is dry and corky. Then the deeper layer of the skin is more oily, with an aroma which the ticks appear to dislike. . . .[54]

Based on Hunter's theory, then this ability to withstand heat seems to be primarily related to the insulating qualities of the Brahman's hide.

Another popular hypothesis used to explain the ability of Brahmans to withstand heat has to do with the actual surface area of their hides. Extra skin in the dewlap and sheath and over

One popular hypothesis regarding the Brahman's ability to withstand heat extremes has to do with the actual surface area of its hide. With the extra skin in the dewlap and over other parts of the body, there is provided a greater amount of surface area than other breeds per unit of body weight.

other parts of the body provide a greater amount of surface area per unit of body weight.[55]

Dr. Turner says that he feels most research up to this time on why Brahman cattle are able to withstand heat doesn't really prove that there is any particular aspect of their nature that make it possible. He feels that over a period of time *Bos indicus* cattle developed "certain genetic entities that gave them the ability to withstand extremes of heat . . . and trying to explain this particular adaptation is a jungle." The important thing is the fact that this trait can be passed on from generation to generation or when Brahmans are crossed with *Bos taurus* breeds.[56]

Many of the early investigators of this question thought that color was the greatest factor in this heat tolerance. J.W. Sartwelle, in his early observations said, "Another remarkable peculiarity of the Indian breeds is the fact that, however white the hair, all but a very small percentage have jet black skins underneath. . . . There is little doubt that the black skin has much to do with the ability of Indian cattle to work in the sun without suffering as light-skinned cattle do."[57]

Considering the outer hair, Turner suggests that color is a characteristic that cannot always be related to heat adaptability; however, he adds, "From my own observations, light colored skinned cattle seem to be less bothered by direct sunlight."[58]

In studies made during the 1960's, results pointed out that Brahman cattle had higher red blood cell counts, total cell volume and hemoglobin values than did Herefords under heat stress. The venous blood of Brahman cattle were shown to have less carbon dioxide than their Hereford counterparts. "These results implied that Brahman cattle are capable of maintaining lower respiration rates during periods of high ambient temperatures."[59]

Data accumulated at the Iberia Livestock Experiment Farm at Jeanerette, Louisiana substantiated these findings. "Respirations per minute by European cattle were found to be 28 in the shade, but when exposed to the summer sun, they increased almost 500 percent-to more than 130 per minute." On the other hand, the rate in purebred Brahmans varied by only 12 from a low of 20 in the shade to a high of only 32 per minute when exposed to the sun. Brahmans in this study subsequently showed a smaller increase in body temperatures. While the temperatures in European cattle increased approximately three degrees when they moved from the shade to the sun, the temperatures of the purebred Brahmans increased only three-fifths of one degree during this same change.

In these same studies a definite correlation was also established between the amount of Brahman blood cattle had and the

A number of state universities now have their own experimental Brahman herds. This particular herd is owned by Mississippi State University at Starkville, Mississippi. Brahman breeders have helped to establish most of these herds.

degree of heat tolerance they had. In other words, as Brahman blood was increased, heat tolerance was correspondingly increased or improved. For instance, half-blooded Brahmans faired better than quarter-blooded Brahmans during these heat tolerance studies.[60]

In still another study in which skin temperatures and respiration rates between Brahman and Jersey cattle were compared at air temperatures of 50°F and 75°F, it was found by T.E. Allen that Brahman cattle have lower respiration rates at all levels of skin temperature, and skin temperatures of Brahman closely followed air temperature. In considering the significance of these findings, "It should be noted that Jersey cattle are generally accepted as adapted to warm climates."[61]

Investigative work by T.C. Cartwright in 1955, in which "summer gains" of Brahman and F_1 crosses and Hereford cattle were compared, a marked advantage was observed in the Brahman and F_1 crosses.[62]

One of the reasons that Brahman purebreds and Brahman crossbreds gain so much during the summer and in warmer climates is due to their insensitivity to direct sunlight. One only has to observe a Brahman herd for one hot, sunny day to see how oblivious they are to the heat and sun. While other breeds will seek shade, Brahmans will continue to feed. In his studies at the U.S.D.A. Iberia Livestock Experiment Farm, Dr. A.O. Rhoad found "that purebred Brahmans spend close to 100 percent of the daylight hours either grazing or resting in the sun, where Aberdeen-Angus, in Louisiana, will spend 40 percent of their time resting in shade, and grazing only 28 percent as against the Brahman's 43 percent."[63]

It has been demonstrated time after time that as the weather becomes progressively warmer the condition of appetite and other factors of general health will decrease the efficiency of feed use on the part of European cattle. On the other hand, unless the weather is extremely hot, it will have no adverse effect on Brahmans.[64]

This ability to fatten Brahmans and Brahman blooded cattle during the summer months is an important economic factor, particularly in the warmer climates. Because of this quality, feeders are able to make year round use of their feed lot facilities; and "Less labor is required to feed in summer than in winter because cattle may be held in pens rather than housed in barns. Brahman steers are admirably equipped for grassland farming because they can turn cultivated grass pastures into immediate profit."[65]

Higher Carcass Yields

Besides R.B. Kelly's investigative work in Australia on heat tolerance and tick resistance, he also completed studies on weights and carcass types of Brahman-crosses. Using herds of 22 half-bred and 57 quarter-bred Brahmans, Kelly recorded that

Studies by Dr. A. O. Rhoad at the U.S.D.A. Iberia Livestock Experiment Farm demonstrated that purebred Brahmans spend close to 100 percent of the daylight hours either grazing or resting in the sun. (This photo was made on a Guatemalan ranch.)

their "average dressed weight was 688 pounds, and they were graded as 86.9 percent *first* grade." These were particularly impressive figures as it was considered then (1940) that under existing conditions in Central and Northern Queensland, Australia, first quality grading above 80 percent was regarded as meritorious. Kelly said, "Although the hybrids appear 'leggy,' the carcasses have thick and well rounded hind legs, and have been described as indistinguishable from those of British bred cattle, and admirably suited to trade requirements."[66]

Commenting on a four-year-old Brahman-Shorthorn he slaughtered at the Brisbane abattoir, a "master butcher reported that he had never broken up a better hindquarter 'as it boned out marvellously in that it had no waste, and that in quality and salable beef it was far in advance of those available to him in the Sydney meat trade'."[67]

One of the most extensive early studies conducted on carcass yields for Brahman crossbred steers was completed at the Texas A & M College Meat Laboratory during the 'Forties. Here large representative samples of carcasses from British breeds and Brahman-Hereford crosses were broken down into standard wholesale cuts and evaluated as to their relative consumer worth. "There was no difference in the U.S. Market Grades assigned the carcasses of each type ranging from U.S. Low Good to U.S. Low Choice, but the rib-eye area of the Brahman crossbred steers was larger than that of the English type steers, and the Brahman crossbreds had heavier loins and rounds.* And it should also be pointed out that the Brahman crossbreds were generally weaned two weeks earlier than the British steers and maintained a weight advantage until they were slaughtered; consequently, yielding more salable meat after butchering.

A summary of these carcass studies showed that Brahman crossbreds gave an overall 3% higher chilled carcass yield than did British breeds.[68]

Carcass yield studies were also completed at the Iberia Livestock Experimental Farms at Jeanerette, Louisiana from 1943 to 1949. Brahman and Angus crossbreds were compared with purebred Angus cattle. The half-breds excelled in carcass yield, and purebred Angus required approximately 92 to 100 days longer than Brahman-half-breds in reaching slaughter weights and approximately 65 days longer than Brahman-quarter-breds.

Tests at Jeanerette, Louisiana also revealed that Brahmans and Brahman-crosses topped all other cattle in percentage of

*The round, with rump and shank off, of the Brahman crosses in these studies outweighed Hereford steers by a pound. The rib was half a pound heavier on the crossbreds, and rib-eye tracing on Brahman crossbred catttle measured 9.8 sq. inches as compared to 8.6 sq. inches in Hereford carcasses.

Brahman-cross cattle usually produce outstanding carcasses. This carcass was the Grand Champion carcass at the San Antonio Livestock Exposition in 1964. It is a Brahman-Polled Hereford-cross and was bred by Butch Roberson of Devine, Texas and shown by Clinton Boerner of Comfort, Texas.

time used for grazing. During the daylight hours, they rested in the shade only 3% of the time.[69]

Investigative work at the University of California on the dressing percentage of English bred steers and Brahman crossbred steers in the 1950's showed a $10 to $12 advantage per head for the Brahman-crosses at 1,000 pounds live weight.[70]

A popular misconception that has been rejected by research is that breed effects taste. According to Dr. G.T. King, Department of Animal Science at Texas A & M University:

> In all of the consumer preference work I have done in the past two years . . . to a panel of 40 families, not one family was able to detect the difference in meat from purebred Brahman cattle from that of Hereford, Angus, Shorthorn, Charolais and Santa Gertrudis . . . . rated in some instances it was rated higher than meat from the above mentioned breeds.[71]

Weight Gains, Survival and Mortality

At one Texas pasture station a number of early studies were made on rates of weight gain comparing Brahman-Hereford crosses with grade Herefords. It was demonstrated clearly during this research that crossbred calves mothered by half-blood Brahman-Hereford cows have an advantage over calves mothered by Hereford cows and sired by half-blood Brahman-Hereford bulls. After seven months, calves from Hereford cows averaged 334 pounds as compared to 469 pounds from half-Brahman and quarter-Brahman cows. Other studies at Lufkin also revealed that Brahman-Hereford crosses demonstrated greater longevity. After nine years most of the grade-Herefords had died while nearly 70% of the Brahman-Hereford crosses were still alive and healthy. Kelly's studies in Australia also indicated that mortality rates among Brahman crossbreds was 30% lower than among graded British breeds.[72]

In 1942, F. H. Finch, manager of Estancia Garruchos, Provincia de Corrientes, Argentina, purchased 10 cows and two bulls from the Sartwelle Brothers of Palacios, Texas. "All ten of the original cows were still living at past 15 years of age, in 1954." Other records revealed that "in spite of the roughest of environmental conditions the 10 American born Brahman cows had given an average calf crop of 80 percent. Two of the cows had calved every year," Finch reported in 1954. He also said that ". . . with the use of American Brahman bulls he had reduced the mortality rate among his cow herds from a high of 14 percent annually to less than one percent." He stated further that he had increased his calf crop from 45 percent up to 80 percent by 1954 as he increased the amount of Brahman blood in his herds. This ranch was running 10,000 cattle annually.[73]

American Brahmans introduced into the research herd at the Cooperative Agricultural Experiment at Tingo Maria, Peru beginning in 1938 also caused a dramatic effect on existing mortality and survival rates.[74]

Established through the cooperative efforts of the Peruvian and U.S. Governments, the Tingo Maria station is located in a tropical area some 900 miles south of the equator at about 2200 feet above sea level. Until American Brahmans were brought from Texas to the station, the few hardy farmers who had tried to raise beef and dairy cattle had had practically no success. Ticks, other insects, and the heat had wiped out nearly all the cattle which at that time were *Bos taurus* breeds.[75]

When a beef cattle program was eventually started at Tingo Maria, carefully kept records showed that over a period of 8 years the cattle lost were mostly pure *Bos Taurus* cattle. Of 79 that died during this time, less than 10 carried Brahman blood. When Brahmans were crossed with Herefords, the progeny also demonstrated a superiority in growth rates, total growth, disease and insect resistance, heat tolerance and milk production.[76]

Brahman Blood Boosts Dairy Breeds

While most of the early research projects on Brahmans were established to answer questions about beef production, a number of early programs were organized for investigative work

In Latin America, Brahman-blooded cattle are often used to upgrade dairy cattle. In some areas, cattle simply do not survive without an infusion of Brahman blood. These cattle are Brahmans crossed with Criollo cattle in El Salvador.

on dairy cattle. One of the first of these studies took place in Jamaica in 1910 and though it did not include what we now know as the American Brahman, it did involve *Bos indicus* cattle.

Directors of the Hope, Jamaica Research Station concluded that the injection of high quality Zebu blood into dairy herds would be necessary before desirable dairy cows could be developed in this tropical region. Three breeds of *Bos indicus* were tried, the Nellore, the Hissar and the Sahiwal. The Sahiwal, or Montgomery as it is often called, had long been know as a milk cow, and subsequent studies indicated it had more potential than the other two. It was crossed with Guernsey, Holstein and Jersey breeds.

These Sahiwal crosses, like other *Bos indicus*-crosses, produced cattle resistant to most tropical complaints and were particularly resistant to tick fever. J.W. Howe also concluded in 1948 that animals with Brahman blood had "a higher birth weight and grew more rapidly than northern purebreds, the rate of growth increasing with the amount of Brahman blood. . . . The age at which heifers first calved also increased with the amount of Zebu blood, as did length of dry periods, length of service periods, and the gestation period."

It was also observed that milk production during lactation increased as the amount of Brahman blood increased with the greatest production being given by half-breds. And butterfat content of milk increased with Brahman-Jersey crosses approximately one percent over purebreds.[77]

In a study held at Angleton, Texas from 1962 to 1967 on the ability of various breeds to produce milk, the superb milk producing qualities of the American Brahman were strongly shown. This research initially involved the use of 89 European cows and 46 Brahman cows and showed a 26 percent advantage for the Brahmans. Additional studies involving the use of 33 Brahmans and 58 European cows gave the Brahmans a 22 percent advantage during early lactation, and during the "critical late lactation period this advantage increased to 44 percent."[78]

As American Brahmans were imported into Australia, a number of these breeds were also crossed with native dairy herds, made up principally of British breeds. Some of the dairy herds developed from these unions have given ". . . as much as 20% more milk than herds of European breeds under the same conditions. Moreover, they responded more quickly to improved conditions in their natural feed. . . . Brahman-crosses have also been found to withstand prolonged droughts better than European purebreds. Other desirable features . . . have been resistance to ticks, (other) insects, cattle diseases generally and (increased) longevity."[79]

Dr. Ralph W. Phillips, one of the world's greatest authorities on tropical livestock, and a group of United Nations scientists analyzed the milk production records of some 1,187 Brahman blooded dairy cows from Jamaica and Trinidad. The results of their study showed that milk production increased as the amounts of Brahman blood increased, the greatest production being given by half-breds. They also concluded ". . . that animals with Zebu blood have a higher birth weight and grow more rapidly than purebred animals of European breeds . . . . The amount of growth rate is more perceptable after two years, and

Noted for their spotted coloring and "buffalo-type" horns, Gir cattle have the reputation of being good dairy cattle. The Gir breed was also used in the development of the American Brahman. (Courtesy Dr. Frank D. Yturria, *Bos indicus, A Brief History . . . of some of the Different Breeds.)*

at maturity half-breds weighed, on the average, 179 pounds more per head than purebreds."[80]

Other significant projects for developing Brahman-crossbred dairy types have been established in the Philippines, Brazil, Peru, Cuba, India, French West Africa, Tunisia, etc. In these areas, the Brahman crossbreds have a high survival rate, and do a fair job of producing milk. While better dairy cattle might exist in other parts of the world, in many of these tropical and subtropical regions, dairy cattle could simply not survive without Brahman blood.[81]

Of the four *Bos indicus* breeds that have contributed to the American Brahman's genetic pool, the Gir has the reputation of being the best milk producer. Because of this quality, it has gained some popularity in the United States although there are no large herds of pure Gir here. In India the Nellore cow, "regarded as the poor man's cow," is also efficiently used for milk production. While the American Brahman itself is a beef breed, many are now being used successfully around the world on dairy farms to upgrade existing herds.[82]

Brahman Breeders Provide Brahmans

Much of the research and investigative work on Brahmans has been initiated by Brahman breeders. On a number of occasions American Brahman breeders have provided experimental stations and universities with Brahman cattle on a loan basis or as an outright gift for the purpose of experimental work. For instance, the J.D. Hudgins Ranch loaned a bull to the East Texas Pasture Station at Lufkin, Texas in 1940, and in 1947, they established a purebred American Brahman herd on the campus of Texas A & M for educational purposes. "This was a replacement loan, where the original female offspring from each of the cows was kept by the college."[83] One year later, the following ABBA members loaned purebred Brahmans to the Bluebonnet Farms Experiment Station at MacGregor, Texas; L.M. Slone, Burke Brothers, E.W. Boyt and C.K. Boyt, Stewart Savage and Sons, J. T. Garrett, Raleigh Sanborn, Henry Koontz, Robert C. Dunks, Steinmann Cattle Company, Malcolm Levi, A. G. Dingle, Wilbourn S. Gibbs, F. W. Neuhaus and J. D. Hudgins Ranch.

In 1953 this research herd was enlarged with purebred Brahmans being loaned by C.K. and E.W. Boyt, J.D. Hudgins, Inc., V-8 Ranch, Vernon Frost, Gail Whitcomb, Guy L. Jeanes, Jr., J.T. White, G.A. Parr, C.Y. Jacobs, C.E. Yoakam and Albert B. Fay.[84]

The Inter-American Institute of Agricultural Sciences of Turrilba, Costa Rica received five registered American Brahman heifers during the 1950's that were donated by C.E. Yoakam, Vernon Frost, J.D. Hudgins Ranch and the Pinehurst Farms of Orange, Texas.

In 1954 the University of Missouri was loaned American Brahman calves to be raised in its climate laboratory where San-

These American Brahmans, known as Barzee Brahmans, were loaned by J. V. Gates to the University of Missouri in 1947 for experimental work.

ta Gertrudis and Shorthorn calves were also being raised. Loaning these calves were Gail Whitcomb, Wilbourn Gibbs, the V-8 Ranch, C.Y. Jacobs and Southern Rice Farms, Inc. of Carlisle, Arkansas.

Texas Technological College at Lubbock, Texas was provided a herd of registered American Brahman females in 1951 by A.P. George, C.K. Boyt, J.D. Hudgins Ranch, O.J. Flowers, J.W. Pate, the B. H. Willis Estate, Henry Koontz, Gail Whitcomb, and H. T. Stock Farms of Natchitoches, Louisiana.

In 1952 Bob Evans of Galliopolis, Ohio loaned an American Brahman bull to the Carpenter Experiment Station of Ohio State University where experiments were being run to determine the dollar value of calves at weaning. Of the breeds investigated, the Brahman-crossbred averaged the highest. A year later, the

Eastern Brahman Association gave the University of Florida ten purebred American Brahman heifers.[85]

A Breed For The Future

Research involving the American Brahman has made a number of notable contributions to the world's livestock industry. Importantly, it has helped to introduce the Brahman to many who were unaware of its extraordinary qualities, not the least of which are its great versatility and endurance. While the vast amount of evidence from this research has been predominatly complimentary of the Brahman and has reaffirmed what its advocates already knew, some problems have also been documented and have been focalized for Brahman breeders and animal scientists to set right. New or proven management techniques and different crossbreeding programs developed on the ranges and at the experimental stations have already corrected or reduced a number of these problems. Others will have to be solved in the future. Above all, research over the past half-century has shown "that the American Brahman, above all other breeds, possesses the capacity for improvement within itself."[86]

Overall studies have shown that from a physiological standpoint, including "hair color, skin color, ability to walk, intelligence, the ability of self preservation, ability to reproduce, livability, coating characteristics, skin thickness, feeding and grazing habits that the American Brahman is truly the formula for efficiency in beef production . . . ."[86] While the American Brahman has not yet scored a perfect grade, it has certainly passed all tests with flying colors and has established itself in the minds of many cattlemen as the breed of the future.

7
Brahman, Crossbreeding's Common Denominator

Since their introduction into North America over 140 years ago, *Bos indicus* cattle have been used more extensively for crossbreeding than any other cattle breed. The results have been phenomenal with "the offspring having the vitality and vigor of the Brahman and the additional beefiness of the British breeds." While crossbreeding among different European breeds has also been pursued, it has seldom produced such dramatic results without the addition of Brahman blood. It has been the genetic diversity between Brahman and European breeds which has maximized heterosis. "When Brahmans are mated to European breeds, the resulting offspring has exhibited the zenith of hybrid vigor in beef cattle," according to cattleman Leon Locke, a former president of ABBA.[1]

Generally, Brahman crossbreeds are "sleek, short haired animals with loose mellow hides," and they have established a reputation for finishing out rapidly both in the feed lot and on the range. Most of these hybrids have been marketed as calves, as they are ready for market earlier than the average breed. By shipping the halfbred calves earlier, the dams have also had a chance to fatten and be in condition to bring a good calf the successive year.[2]

One of the earliest, popular Brahman-crosses was the Brahman-Shorthorn cross. Pictured here are F_1 Brahman-Shorthorn heifer calves at weaning. Notice the "sleek, loose, mellow hides."

The Brahman-European Derivative Breeds

Santa Gertrudis

One of many eventual products resulting from crossing Brahman cattle with British breeds has been the so-called derivative breeds. In order to establish a particular breed for a particular environment with a certain uniformity of color and conformation, some breeders eventually "carried crosses on further by breeding hybrid to hybrid; that is, cross-bred to cross-bred." In essence they have succeeded in establishing a group of new breeds. The first of these new derivative breeds to be established in the United States was the Santa Gertrudis.

Its story began in 1910 when Thomas M. O'Connor gave the great and famous King Ranch of Kingsville, Texas one of his large, black half-Brahman and half-Shorthorn bulls. Placed in a pasture next to some 3,000 purebred, but unregistered Shorthorns, the O'Connor bull jumped the fence and became responsible for the unplanned pregnancy of a number of Shorthorn heifers.

All of the bulls from the matings were eventually castrated with the exception of an exceptional bull named *Chemmera*. The female offspring were put into a pasture with a number of purebred Shorthorn bulls. But little attention was paid to any of the crosses until 1918 when about sixty of the best descendants of the amorous O'Connor bull and of his son *Chemmera* were turned out into a good pasture.

While the management of the King Ranch had labored for some forty years to establish purebred herds, they became convinced at this time that a Brahman-Shorthorn cow was ideally suited to their range; so they set about to develop a new breed. The owners of the King Ranch, the Robert J. Klebergs, were very familiar with the development of the Shorthorn in England, and they were aware of the fact the *Bos indicus* cattle had been bred for a much longer time, and that their influence had been felt in many parts of the world. "Why couldn't a new breed be developed for their own range?" Since they were so convinced of the superiority of their new crossbreds, they initiated a program involving all purebred Shorthorns. As one reporter later noted, "With the help of his brother, Richard, Bob Kleberg set out to invent his own cow."[3] And with the help of a group of animal geneticists and a veterinarian named J.K. Northway, he did just that.

The real beginning of the Santa Gertrudis was the now famous bull named *Monkey*, so named because he was so playful, whose prepotency served to stamp the characteristics of the breed on his offspring.* This calf was branded in 1920 and "was

*Interestingly, *Monkey* almost became a steer in 1920, when Richard and Bob Kleberg, Jr. argued over whether or not the young bull should be castrated. (See William Broyles, Jr. "The Last Empire," *Texas Monthly*, Oct., 1980, p. 250).

The Santa Gertrudis, developed by the King Ranch of Texas, is well adapted to the semi-tropical climate of the coastal range area, but it has also been a popular export animal, particularly to South America. This breed is approximately three-eights Brahman and five-eights Shorthorn. The famous bull, *Monkey*, **is considered the patriarch of the breed.**

Courtesy Gemini Springs Farm; Debary, Florida

essentially a first cross of Brahman and Shorthorn, both parents being somewhat impure as to breed." *Monkey's* sire was *Vinotero* of the Guzerat type but he also reportedly showed considerable Nellore, Krishna Valley and some Sindhi characteristics. His dam possessed about one-sixteenth Brahman blood which had come down from the original O'Connor bull.

By using *Monkey's* sons and grandsons on first cross heifers, and again on the double cross, resulting from mating first cross bulls on first cross heifers, and finally adopting in-and-in line breeding methods, the Santa Gertrudis breed evolved.* Today their blood contains approximately three-eighths Brahman and five-eighths Shorthorn. Commenting on the future of crossbred cattle with a Brahman base, W.H. Black of the USDA said in 1936 that the development of the Santa Gertrudis crossbred cattle by Kleberg was an outstanding accomplishment of beef cattle breeding along this line.[4]

The breed is red in color with limited white spotting only being permissible on the underline. It is moderate to large in size with mature bulls in pasture conditions weighing from 1700 to 2200 pounds. Under similar conditions, cows may weigh up to 1300 pounds.

News of Kelberg's new breed was soon generally circulated, and in 1943 the wealthy William duPont of Wilmington, Deleware visited the King Ranch hoping to buy some of Kleberg's "new breed" females with the purpose of establishing a Santa Gertrudis herd back in Virginia. As Kleberg was attempting to stock the ranch then, he refused to sell any of the females. But he did offer to loan duPont some excellent Santa Gertrudis bulls and suggested that he develop his own herd.

Following his advice, duPont contacted the J.D. Hudgins Ranch where he worked out the lease of two top *Manso* Brahman sires for a 3-month period. Edgar Hudgins later put him in touch with John Impson of Beeville, Texas who owned one of the best Shorthorn herds in the United States.

DuPont eventually acquired 80 half-*Manso*, Brahman-Shorthorn females from the Hudgins' Brahman bulls and the Impson Shorthorn cows. These females were shipped to duPont's Virginia ranch at Boyce, Virginia where they were mated with some of the King Ranch Santa Gertrudis sires. In time duPont established what Bob Kleberg called, "one of the finest graded-up herds of cattle in the world," and when William du-

*Kleberg named his new breed after the small creek located near the spot where Richard King started his first rancho in 1853. (See William Broyles, "The Last Empire," op. cit.)

Pont, Jr. died in 1966, the King Ranch bought this herd by sealed bid.

Kleberg was so impressed with these cattle that he set about to introduce two-thirds of the purebreds into the King Ranch foundation herd as an outcross. "This was the first time in the history of the breed that outside female bloodlines had entered the King Ranch purebred Santa Gertrudis program since the death of *Monkey* . . . ." The remaining one-third of the duPont Santa Gertrudis herd was sent to the Big "B" Cattle Company of Florida, "a joint venture partner" with the King Ranch.[5]

Since the establishment of the Santa Gertrudis breed, an association was founded in 1951 named the Santa Gertrudis Breeders International. And today, the breed is utilized in forty-nine states of the United States and in approximately sixty countries. It has become an important export animal and is noted for its adaptability for tropical and semi-tropical areas.[6]

Beefmaster

Chronologically, the next Brahman-European derivative established in the United States was the now popular Beefmaster. The story of the first Beefmaster foundation herd actually begins in 1908 when Ed C. Lasater of Falfurrias, Texas began introducing Brahman blooded cattle, presumedly from Borden stock, into his Shorthorn and Hereford herds. He also later purchased a number of cattle from the 1925 Brazilian Importation that came via Mexico. (See Chapter IV, "Those Brazilian Brahmans.") "Special emphasis was placed on developing a strain of cattle that would produce a choice, quick maturing, heavy calf at eight months of age, under range conditions with no supplemental feeding."

In 1931, Tom Lasater, Ed Lasater's son, began a crossbreeding program using his father's registered Hereford and Brahman herds along with some registered Shorthorns. It was from this three-way cross, which was made up of about 50 percent Brahman hereditary material, that the Beefmaster was actually developed.

During the 1930's, Lasater closed the foundation herd to outside influence. Once his original blend was made, it is said that his culling was ruthless.*

South Africa's Dr. Jan C. Bonsna is said to have labeled the Lasater Beefmaster herd as ". . . near perfection in functional efficiency." And Tom Lasater credits the Brahman blood for a good share of the herd's achievement.[7]

Because the Beefmasters are bred under range conditions, no exact pedigrees are kept, but it is believed that besides the 50% Brahman blood, the remaining percent is divided about equally between the Shorthorn and the Hereford. Color is normally disregarded, but as the herd has grown, each year a higher percentage of calves have been red.

Under warm range conditions, the Beefmaster will reportedly outweigh European breeds by 30% at any age. Its high percentage of Brahman blood has also given it a high resistance to heat, disease and insects as well as giving it good rustling ability.[8]

In tests conducted by Texas A & M College at Balmorhea, Texas in the 1940's, ". . . a group of six Beefmasters made an average daily gain of 2.5 pounds during the 143 days of tests." These tests also included other breeds, but the two top groups both carried Brahman blood.[9]

The Beefmaster has demonstrated the ability to adjust to a wide range of climatic conditions. Although it was developed in Southeast Texas where the climate is sub-tropical, it has faired well on open lots where the temperature may commonly drop to

*"Only descendants of the Lasater Beefmasters, or cattle produced by three consecutive top crosses of recognized Beefmaster breeding, may be called Beefmasters," Each animal must also be identified with a proper prefix, such as Lasater Beefmasters, Casey Beefmasters, etc., so that each breeder is responsible for his own product.

There are actually three Beefmaster breed associations. One is the Foundation Beefmaster Association with headquarters at Denver, Colorado. The others are the Beefmaster Breeders Universal located at San Antonio, Texas and the National Beefmaster Breeds, Fort Worth, Texas.

Courtsey Southeastern Beefmaster Breeders Association

The Beefmaster, developed from a three-way cross, made up of about 50 percent Brahman with the rest Hereford and Shorthorn, has become well known for its functional efficiency. Like other breeds developed from Brahman stock, the Beefmasters are known for their great versatility.

23° F. below zero. In fact, Tom Lasater eventually moved his foundation herd of Beefmasters to a range near Matheson, Colorado in 1949.[10]

Brangus

In an attempt to utilize the extraordinary capacities of the American Brahman with the prepotency for fertility, carcass

Coallie, **pictured above, is the first Brahman bull used with Angus cows in crossbreeding studies at USDA's Iberia livestock experiment station, Jeanerette, Louisiana, starting in 1930. These Brahman-Angus crosses initiated the development of the Brangus breed.** *Coallie* **was from the Dr. W. S. Jacobs' herd at Houston, Texas.**

quality and general type of the Angus breed, interest in developing a Brahman-Angus breed existed for many years. This interest was greatly increased during the late 1930's and 1940's at the Iberia Livestock Experimental Farm at Jennerette, Louisiana when Brahman bulls from the Jacobs and Hudgins Ranches were crossed with Angus cows.* Comparisons of weights between European cattle and these new crosses showed a decided advantage for the Brahman-Angus cattle, running as high as 272 pounds. When slaughter time for purebred Angus was com-

*Efforts had been made since 1914 at the Iberia Livestock Experimental Farm to develop livestock adaptable to the Gulf Coastal area.

Brahman-Angus crosses in a Texas feedlot. Such feedlot cattle have historically shown an outstanding growth rate, good grading qualities of carcass and moderately high yield of trimmed retail cuts.

pared to these crosses, it was shown that the Angus calves took 92 to 100 days longer to reach slaughter weights. It was also demonstrated that these Brahman-Angus-crosses topped all other European cattle in the percentage of time used for grazing.[11]

From this Brahman and Angus cross, the Brangus breed was eventually developed with a three-eighths Brahman and a five-eighths Angus combination. Actually in the early days when most breeders were starting from scratch, there were a lot of one-half and one-quarter blood Angus and Brahmans, but the major problem in breeding one-half and one-quarter blood turned out to be one of management, resulting from running three or more different herds. While pure Brangus females are available in many parts of the country today, they are relatively scarce in the middle West. And there still some "purist" breeders who like to start a Brangus herd from a Brahman cow base.[12]

One of the advantages of using the one-quarter times one-half route has been to intensify a particular line of Angus a breeder may have in his herd. Brahman bulls can be used on a set of Angus cows and then come back over the one-half bloods with an Angus bull or bulls of the same family. This gives a breeder a set of one-quarter blood cows "which are good 'mamas' and particularly well suited for Oklahoma or the Middle West." In this case one will also end up with one-quarter-blood bulls left over for surplus.[13]

Another formula was started by Phil Jenkins of Spearman, Texas in about 1950. Jenkins started with a group of high quality Brahman cows bred to the best Angus bull he could find in Texas' Panhandle. When his one-half-blood cows were ready to be bred, he put their "daddy" back over them. Jenkins later obtained a high quality one-half-blood bull which he bred to his one-quarter-blood females. "The first of the Brangus calves from this cross were ready to be inspected by about 1961 or 1962 . . ." Although it took a few years to get this program under way, Jenkins demonstrated one way that a good, small breeder could get started in the Brangus business.[14]

While various programs have been used to produce the three-eighths and five-eighths combination, over 85 percent of the 150,000 registered (1976), were the result of Brangus to Brangus mating.[15]

In 1949, interested breeders formed a Brangus organization that later became the International Brangus Breeders Association located at San Antonio, Texas. "The breed registry has remained open and by mating three-fourths Brahman and one-fourth Angus bulls to Angus cows (generally referred to as the shortcut method) the progeny that are approved may be added immediately to the breed registry."

Much of the early breeding in actually establishing the Brangus breed was done by Raymond Pope, manager for and partner of

The Brangus breed was first developed by combining Brahman and Angus breeding in a three-eights - five-eights combination and selecting for the desired traits of both breeds. Registered Brangus must be solid black in color, polled and approved by an International Brangus Breeders Association inspector.

Frank Buttram of Welch, Oklahoma. Before moving to Oklahoma, Pope had worked on a ranch near Franklin and Jeanerette, Louisiana with a crossbred program involving Angus and *Manso* Brahman bulls from J.D. Hudgins.[16]

One of the earliest problems that faced Brangus breeders was maintaining the black color of the Angus they wanted to keep. They soon learned that in order to obtain a beefy, black offspring, it was necessary to use a light gray American Brahman. Both dark Brahman cattle and many Angus cattle, on the other hand, carry recessive red genes increasing the chances of red progeny.

Brangus females are noted for superior maternal ability and good fertility, and Brangus bulls in recent years have worked well in crossbreeding with cattle principally of Hereford breeding in the Southwestern United States. Today Texas ranks as the number one Brangus state.[17]

Braford

Since Brahmans were first crossed, the Brahman-Hereford have been one of the most popular crosses in the United States.

Part of this has been because of the wide distribution of Herefords. First introduced into the United States in 1817, at one time grade commercial Hereford cattle constituted more than 50% of the entire beef cattle population of this country. Surprising though is the fact that it was such a long time before anyone developed the Braford, a new breed, developed from the combination of the Brahman and the Hereford.[18]

Braford cattle are approximately three-eighths Brahman and five-eighths Hereford. "Superior Hereford sired bull progeny from a Brahman-Hereford sire rotation (crisscross), were mated to females of the same breed composition to form first generation Brafords. These cattle have been straight bred for three to seven generations."

The production traits "that have been emphasized in selection have included fertility, calving ease, calf survival, weaning weight and adaptability to their environment."[19]

After sensational progress had been made through an intensive large scale selection program by the Adams Ranch at Fort Pierce, Florida, the International Association of Braford Breeders was formed in 1969. Other American breeders actively engaged in raising Brafords include Francis Savage and Sons (originally Stewart Savage and Sons) of Bay City, Texas and the McGill Brothers of the Alice-Falfurrias, Texas area. In observations made by the McGill Brothers, it was found that their Brafords were especially resistant to *hemmoraghic septacemia* and less susceptible to anthrax, blackleg and other ailments common to Gulf Coastal cattle. And they were also found to be excellent rustlers. "At weaning, Brafords will outweigh Herefords from 100 to 200 pounds in warm coastal areas." With Brafords the sires can be bred to either Brahman or Hereford dams, ordinarily giving a good marketable calf. Braford breeders claim that their cattle get as fat on good range as most of the average run of coastal cattle will on feed. Breeders Savage and Sons have adopted the slogan, "Brafords, the cattle that turn grass into fat."[20]

The Australians have also had considerable success with Brafords developed from the Brahman-*Manso* line. They have

Courtesy Texas A&M University

One of the most popular Brahman-crosses has been the Brahman-Hereford cross. This Brahman-Hereford F_1 cow was seventeen years old when this picture was taken at the Texas A&M University East Texas Pasture Station at Lufkin, Texas. The calf she is nursing is her fifteenth.

their own Braford association and spell Braford with an umlaut, thus it is Bräford to the Aussies. Sam Atkinson, noted Brahman breeder and owner of the Bundarra Station in Queensland, recently told Wendell Schronk, ABBA's Executive Vice-president, "The Brahman breed has expanded extremely well in Queensland and other areas of Australia, and with the development of the Braford and Droughtmaster (a Brahman-Shorthorn cross), it has been a profitable venture."[21]

Courtsey Schearbrook Land and Livestock Company

During recent years the Charbray (Brahman-Charolais cross) has been exported principally for use in tropical areas where pure Charolais are not well adapted.

Charbray

When Brahman females were used extensively as foundation females in the process of upgrading Charolais, "crosses of these two breeds attracted wide attention because of their tremendous rate of growth and size." Eventually, the term *Charbray* was copyrighted to include Brahman-Charolais-crosses containing one-half to one-sixteenth Brahman. J.A. Lawton, Jr., of Sulphur, Louisiana, reported in 1950 that his three-eighths Brahmans and five-eighths Charolais-cross produced an even larger animal than he had been getting from his halfbreeds.[22]

Breeders and promoters of the Charbray have generally stated that the ruggedness of the Brahman combine with "the rapid

The introduction of Brahman blood has given the Charolais much better performance records on tropical and sub-tropical ranges. These are F_1 Brahman-Charolais cows.

growth of the Charolais to produce a beef-type animal that is a good grazer, a fast breeder, and one that is easy to handle." Grown cows weigh from 1700 to 2200 pounds; grown bulls may weigh from 2500 to 3200 pounds. One bull calf receiving special attention reportedly weighed over 1200 pounds at one year of age.[23]

Charbray bulls were once used in the United States as a substitute for Charolais bulls before the latter became plentiful.[24] "During recent years the Charbray has been exported principally for use in the more tropical areas where the pure Charolais is not well adapted."[25] Its Brahman blood, of course, is the important thing that has given the Charbray those characteristics to survive in climates otherwise considered inhospitable to any kind of cattle.

In a far different climate, *Bos indicus* cattle have also been crossbred with Charolais in the dry French Sudan and in French Morocco. "The half-bred Brahman and Charolais and the three-quarter bred are said to be excellent." In Argentina, American Brahmans have also been crossed with Charolais with good results.[26]

Courtesy Pat Corrigan, Vero Beach, Florida

This 16 month old purebred Simbrah, *C Honcho,* **is owned by the Pat Corrigan Ranch of Vero Beach, Florida.**

Simbrah

Described by some of its proponents as "The most exciting new breed on the American scene," the purebred Simbrah generally is three-eighths Brahman and five-eighths Simmental. Simbrahs have thus far demonstrated proven performance in a wide range of environments. Presently the American Simmental Association recognizes what it referred to as a "percentage Simbrah" which has to have a minimum of three-eighths Simmental and one-fourth Brahman with up to three-eighths of another breed or other breeds.[27] Similar to other Brahman-European-derivative breeds, the Simbrah is adaptable to an extremely wide range of conditions and climates.

Charles Campbell, Simbrah breeder from Centreville, Mississippi recently reported that "calves come off of his Simbrah cows 600 to 650 pounds at six to seven months of age." He also said that he had Simbrah calves in a Texas feed lot that gained 3.67 pounds per day that came out after five and a half months averaging 1115 pounds.[28]

Other Brahman-European derivatives have also been developed by crossing Brahmans with Limousins, Shorthorns, and Brown Swiss. "The value of Brahman blood in the beef cow is easily demonstrated in the fact that *every other beef breed, or derivative breed, of American origin was developed through a Brahman cross.*"[29]

Several Brahman and European composites have also arisen in the Caribbean area. One of these is the Jamaica Black, descended from Brahman and Angus crosses. It is very similar to the Brangus in appearance and performance. Another crossbred, the Jamaica Red, was developed by top-crossing Red Polled bulls on Criollo and Brahman females over a long period of time. The Jamaica Hope, on the other hand, is a dairy type animal which includes Brahman, Jersey and Criollo breeding. A breed association was actually established for the Jamaica Hope in 1953, and the herd book was closed that year.[30]

In the American tropics, grade-Brahman and crossbred types of cattle likewise are used extensively in dual purpose enterprises where the cows are both milked and permitted to nurse their calves. "While not regarded as progressive by some, economic studies tend to vindicate the practice."[31]

While the majority of beef cattle in the American tropics are still grade-Brahman or Brahman-Criollo crosses in the process of being further upgraded to Brahman, breeds based on crossbred foundations of the improved European and Brahman breeds are currently being promoted in these areas of the world.[32]

The Unbeatable Cow

All real cattlemen have long been aware of the rich returns that can be produced from the crossing of genetically divergent breeds. It has been proven conclusively that an increase in pro-

Breeders in the Southeast and Southwest United States have had considerable success with the Brahman-Limousin cross. Registered through the North American Limousin Foundation, purebred Brahmousin bulls must have a combined Brahman and Limousin blood totals of ninety-three percent. Females must have an eighty-seven percent or greater Brahman-Limousin blood total. This yearling Brahmousin bull is on La Romana Ranch at Madison, Florida, owned by Bill and Leah Kinner.

duction results when the American Brahman is crossed onto *any* breed or type of cattle in the United States.[33] Assuming conditions are the same, the greatest productivity is normally obtained from the first generation, or what stockmen call the F_1 cross. Thorough research studies have clearly demonstrated that the most dramatic and profitable F_1 cross is produced when high quality Brahman cattle are bred with high quality European cattle. Indeed, with Brahman breeders, the term "F_1 cross applies usually to the offspring produced by purebred Brahman and any purebred European breed." Generally, in cowman's parlance, "F_1 refers more specifically to the progeny of a purebred Brahman and a purebred Hereford or purebred Angus."[34]However, the heterosis that is achieved with crosses of purebred Brahmans and Herefords or Angus, for instance, is also clearly apparent when Brahmans are crossed with Charolais, Limousin, Brown Swiss, Shorthorn, etc. Since Herefords and Angus are number one and number two in terms of cattle registered in the United States, most F_1 progeny have subsequently been one-half Hereford or one-half Angus.[35]

Perhaps no one has been aware of the potential of the F_1 cross longer than Harry Gayden, Executive Secretary Emeritus and honorary Vice-President of the American Brahman Breeders

These Brahman-Hereford crosses clearly demonstrate the gainability of Brahman-crosses on grass. This picture was taken on the Glenn Keeney Ranch.

Association. First as a student in the 1920's and then as a member of the staff at L.S.U., Gayden was fascinated with the dramatic results produced from crossing Brahman cattle with other breeds. When purebred Brahmans were crossed at L.S.U., Gayden became convinced that the F_1 concept had great merit. Gayden was employed by the ABBA in 1948, giving him an opportunity to further promote the advantages of Brahman crossing.[36] In 1948, ABBA's president, Gail Whitcomb and Gayden met with animal scientists at Texas A & M and got the ball rolling on a series of new research projects involving Brahmans and Brahman-crosses. ABBA members provided registered Brahmans and a loan was obtained in 1949 to underwrite part of the research.[37] Two years later, an extensive research program was begun at the Texas Agricultural Experiment Station at McGregor, Texas. (See Chapter VI, "Brahmans Pass All Tests.") Promoted by Gayden and the ABBA, studies were extended over a 14-year-period. During this time a total of 3,145 calved were tested. A summary of the excellent performance of the first cross-Brahman hybrid cow is evident in the table below.[38] (See Figure I.)

"First cross calves nursing straightbred Brahman and Hereford cows exceeded straghtbred calves in a 180-day weaning weight by approximately 13 percent. Backcross calves on F_1 Brahman hybrid dams weaned 16 percent higher. "Other evidence at

Dam	Sire	180-day Weaning Weight (Lbs.)	Percent Advantage	
Hereford	Hereford	384	Base (Lbs.) 380	
Brahman	Brahman	377		
Brahman	Hereford	441	16%	13%
Hereford	Brahman	425	11%	
F_1 B×H	Hereford	446	17%	16%
F_1 B×H	Brahman	437	15%	
F_1 B×H	Santa Gertrudis Charolais Brown Swiss	460	21%	

FIGURE I

hand from these stations is that calves out of half-Brahman-English type crossbred cows wean at a weight 25 percent heavier than pure English calves . . . ."[39] Three breedcross calves produced by Santa Gertrudis, Charolais and Brown Swiss bulls on F_1 hybrid dams exceeded the straighbreds by 21 percent."[40]

In studies later conducted at the Everglades Experiment Station at Belle Glade, Florida from 1960-1967, which included Angus, Brahman, and Hereford purebreds and various crossbreds, the F_1 Brahman hybrid dams were also consistently heavier.

Recent studies have been, and still are being, conducted at the U.S. Meat Animal Research Center at Clay Center, Nebraska, which have involved tests with the Brahman F_1 cow. According to one animal geneticist, "There is no cross that is even close to it. It is an unbeatable cow." A cowboy who handles the cattle at Clay Center was recently quoted as saying that he wished all the cattle there were Brahman-crosses when it came to handling them.[41]

One reason that the F_1 Brahman cow has been able to produce such a superior calf is because she is an outstanding milk producer. "She is a more efficient forager and a better mother. And on top of this, the Brahman F_1 has the longest producing life of any beef cow.[42]

At the Texas Agricultural Experiment Station near Angleton, Texas, tests run from 1957 to 1965 revealed that milk yields for F_1 Brahman hybrids were considerably higher than they were for Brahman and Hereford purebreds. Brahmans, incidentally,

This Certified F₁ Brahman-Hereford heifer has those outstanding characteristics that caused one respected animal geneticist to refer to the F_1 Brahman cow as "the unbeatable cow."

showed the second greatest yields. In a more recent study, 1966-1968, first cross Brahman-Hereford dams exceeded the Herefords in milk production by 80 percent. "In three-breed cross dams, milk production almost doubled."[43]

Studies have also shown that the Brahman F_1 cross is an efficient feed lot animal. At the McGregor, Texas Station, purebred Brahman bulls bred to purebred Hereford dams producd calves that consistently outgained purebred Hereford and Brahman calves. "Results of a five-year experiment at the University of California Imperial Valley Field Station at El Centro, California, also showed that Brahman hybrids consistently gained faster than a straight-bred European breeds in drylot under the hot, desert-like conditions that prevailed." Feeding tests run during the winter at El Centro indicated little, if any, statistical difference between Brahman-crosses and other breeds.[44]

In research completed at the Agricultural Research and Education Center near Quincy, Florida feed, efficiency-tests involving F_1 Brahman-Hereford and Hereford-Angus crosses were conducted from 1967 to 1969. In this study, the F_1 Brahman-Hereford steers required 72 pounds less feed per 100 pounds of gain than the Hereford or Angus steers.[45]

As previously mentioned, crossbreeding has also had its effect on carcass characteristics. A Brahman-Hereford F_1 steer, for instance, had the champion carcass at the prestigious, highly competitive Houston Livestock Show in 1978. In this event the top three steers from every class in the show were judged. "The Brahman half-blood beat them all from the new exotics down through the more traditional." The following year another Brahman F_1 steer finished in the top ten.[46]

One of the results of F_1 Brahman breeding is the development of a more uniform carcass with less globs of fat. And characteristically, crossbreeding has resulted in transferring more uniform marbling than is normally found in the purebred cattle of various breeds. "Studies have also shown that F_1 Brahman-crosses generally produce a carcass in the same U.S. Grade with a ten percent larger rib eye and a five percent heavier loin and round than English cattle of equal breeding quality and raised and fed under identical conditions."[47]

While research on carcasses generally shows that carcasses produced by offspring from English breeds will have more outside fat covering than pure Brahman offspring but less internal fat (kidney, heart and pelvic fat) than the purebred Brahman, when Brahmans are crossed with English cattle, both external and internal fat is reduced.[48]

In research reported by O.D. Butler and Herbert Abraham of the Texas Agricultural Experiment Station and by W.M. Cole of the Tennessee Experiment Station, the percent of fat thickness for the F_1 Brahman cow was shown to be .25 as compared to .34 for the purebred Brahman and .60 and .43 for the English breeds. "The percent of chuck, loin, rib and round was higher for the purebred Brahman and the F_1 Brahman than either of two English breeds studied." Simply stated, "Cattle of Brahman breeding tend to have a higher percentage of edible meat with less excessive fat."[49]

A Certification Program for the F_1 Brahman

Following the tradition of Harry Gayden, ABBA's dedicated and personable Wendell Schronk, has also labored with notable success to popularize Brahman crossbreeding. Schronk recently devised a blueprint for certifying F_1 Brahman females which was endorsed by the ABBA in 1979. This certification program has provided a system for giving deserved recognition to the true F_1 Brahman female as well as serving as a registration plan for identifying the dam's pedigree. While all this is important enough, the greatest vintage to come from this program thus far, has been in providing a common ground from which the various breed associations have been able to develop a mutuality of purpose, direction and understanding.

On May 10, 1980, history in no slight way was made when the cattle industry's first joint Brahman Association and Hereford Association field day was held at the Dudley Brothers Hereford Ranch in Comanche, Texas. While several years of planning and hard work by the American Brahman Breeders Association, the Texas Brahman Association, the Texas Hereford Association and the American Hereford Association were behind this event, the new F_1 certification program was undoubtedly the catalyst that made it all possible. "The common ground for this cooperation was the widespread use and acceptance of the Hereford-Brahman crossbred cow by the commercial cattlemen."[50]

Attended by 600 people, this field day provided the followers of each breed "an educational opportunity to learn more about the other." At this extraordinary event, as Joe Brockett reported, "Brahman breeders were overheard praising the fertili-

Pictured here are some of the organizers of the eventful Brahman-Hereford field day hosted by the Dudley Brothers Hereford Ranch at Comanche, Texas on May 10, 1980. (L to R) James Grote, Llano, Texas; Lovell Kuykendall, American Hereford Association, Kansas City, Mo.; Johnny Summerour, Texas Hereford Association President, Dalhart, Texas; Frank Connell, ABBA Vice-President, Thornton, Texas; Margaret Watkins, Texas Brahman Association; Tom Dean, Texas Brahman Association President; Cherokee, Texas; John Dudley; Jack Chastain, Texas Hereford Association Secretary and Wendell Schronk, ABBA.

ty, rustling ability and freedom from sheath problems of the Horned Hereford breed. Hereford breeders expressed an appreciation for the heat resistance, adaptability and mothering

Dr. Randall Grooms, Texas A&M University, and ABBA's Wendell Schronk, discuss the merits of the Brahman-Angus cross at the first ever joint Brahman-Angus field day, held July 18, 1980 at the USDA Beef Cattle Research Center near Brooksville, Florida.

ability of the Brahman," realizing that all these superior qualities could be found in the crossbred cow.[51]

In June of 1980, a similar Brahman-Hereford field day was held near Montgomery, Alabama sponsored by the American and Alabama Brahman Associations and the American and Alabama Hereford Associations. And in July of that year, a joint Brahman and Angus field day was held at the U.S.D.A. Beef Cattle Research Center at Brooksville, Florida. This event was co-sponsored by the American and Eastern Brahman Associations and the American, Florida and Georgia Angus Associations.

At this latter field day, W.C. "Cal" Burns, Director of the Center, pointed out the advantages of Brahman-Angus crossing, showing several demonstration herds; but he also underscored the importance of selection. As he said, "Mating cull Angus and Brahmans will not make superior F_1 calves." However, as Burns later acknowledged, even when inferior Brahman and Angus cattle are bred, heterosis still takes place, generally producing offspring that excell their parents.[52]

The problem of haphazzardly crossing any kind of cattle for heterosis was demonstrated in the late 1940's and early 1950's. During this time many American cattlemen were just becoming aware of the superior qualities of America's first great exotic -the Brahman. "And cattlemen all over the country were buying anything with a hump." Many of these were poor commercial bulls and did little to help the reputation of the American Brahman. In fact, in some regions it was years before cattlemen would again begin to appreciate the better bred Brahman.[53]

Although the F_1 Brahman Female Certification Program is barely a year old, three unprecedented events have taken place hopefully opening the way to more joint activities among the different breed associations. While there was once an indication that another breed association might institute a second F_1 certification program, it would appear that the ABBA is still the most logical sponsor for this program as the American Brahman is indisputably the "Crossbreeding's Common Denominator." Of course, any other F_1 certification program would also tend to confuse breeders as to which program is the best, thus weakening the whole purpose of the existing program.[54]

Although there is probably no cow that will ever suit all environments and all expectations, one of the real advantages of the Brahman-cross is its adaptability to a wide range of conditions. The potential for producing a cross that can thrive on practically any known rangeland in the world is there - if the right combination is used. While crossbreeding within European breeds has been tried, so little benefit has been derived that the practice has not been so generally accepted. But with the crossbreeding involving *Bos indicus* cattle the results have been phenomenal.

Brahman-European crosses are becoming more and more popular on the Great Plains. These F_1 Brahman-Hereford steers are on a feedlot in Kansas.

Cattle That'll Do The Job

Cowman Dick Pringle of Yates Center, Kansas believes Brahman-crosses are a sound basis for improving herds on the Great Plains of America's heartland. President of the Kansas Livestock Association, Pringle normally runs between 700 and 800 head of cattle during the year, and in the summer, he runs several thousand stocker cattle on native grass. He knows cattle. And he is "very serious about selecting cattle that'll do the job. . . ." Up to 400 head of his cows carry Brahman blood, and he points out that the balance of his herd will be bred back to Brahman-cross bulls or bulls from synthetic Brahman breeds. He also runs several hundred head of F_1 Brahman-cross heifers which are being bred back to British breed bulls.[55]

In the arid and rugged terrain of Hidalgo County, New Mexico, where a cow usually needs 30 or 40 acres of pasture to survive, rancher Bill Cowan runs his Cloverdale Ranch on which he principally has F_1 Brahman-Herefords and registered Brahmans. Currently he is using top herd sires from the A.P. Ranch of Buckeye, Texas, a Paret bred bull, a V8 bull, a 3W bull and one of his own breeding. "In 1974 the National Champion Brahman Cow was owned by the Cowans, a prime example of their high quality herd." Thus far, Cowan's registered Brahmans on his

New Mexico Ranch average a 80-to-85 percent calf crop while his F_1 Brahman-Hereford cows average a 90-percent calf crop - year in and year out. "They also raise a good, fast growing calf that will go into feed lots providing the kind of daily gain and conversion ratios feeders must have." Cowan also has several purebred Angus bulls which he breeds his F_1 heifers to with excellent results.[56]

"Much of the credit for gaining Brahman acceptance in the New Mexico-Arizona area," according to Cowan, "has to be given to Jim Benedict of Benedict Feedyards . . . Jim has fed pen after pen of these Brahman crossbred calves and has really supported them in our area."[57]

Some breeders of F_1 Brahman cattle sell all of their F_1 heifers that are produced each year. One such cattleman is Elgin Guentert of the Bar-D-Ranch near Chappell Hill, Texas. Guentert's F_1 crosses are the progeny of Brahman-Angus breeding, and according to him, the demand for them far exceeds the supply. His F_1 Brahman-Angus steers are usually sold at the auction in Brenham, Texas where they consistently bring a top dollar.

Guentert's total crossbreeding operation is an interesting one, demonstrating again the ingenuity that has characterized the American cowman over the centuries. Although he has held back a few F_1 heifers for replacements to breed back to Brahman bulls, his replacement heifers are usually purebred Angus. "His Brahman bulls are all registered, coming mostly from the top quality lines of J.D. Hudgins, Inc., with some of his bulls coming from Runnells-Pierce."

Guentert says, "I like the uniform color and good mothering and milking qualities of the Angus breed, coupled with the thriftiness and fast growing qualities that the Brahman puts into these F_1 calves."[58]

Another typical Brahman-crossing program involving Brahman and Angus blood is being skillfully administered at the Greenbriar Ranch, located a few miles above the Trinity River bottom in Madison County, Texas. "Purebreds to F_1's to various percentages favoring one breed or the other are produced and

Brahman-Angus crosses are very popular in many parts of the United States. Pictured above are F_1 Brahman-Angus heifers in Southern Mississippi.

marketed here - and even a third breed is brought in to produce a terminal three-way cross."

Four purebred gray Brahman and three purebred red Brahman units and five purebred Angus units are maintained for the primary purpose of producing the breeding material for the multiple crosses. Home-produced Brahman bulls have served a large commercial Angus cow herd to produce Greenbriar's number one project, the F_1 Brahman heifer and steer. Black hybrid steers are sold at weaning time, usually under contract.

Selling F_1 calves by contract has posed somewhat of a problem for Richard Deal, Greenbriar's manager. In order to keep growthy F_1 calves from weighing more than what is stated in the contract, they have to be weaned early. Deal recalls that when one customer specified 500 pound steers, he had to hold his calves back awhile before weighing; thus allowing for a 3-percent shrinkage. "They still wound up 99 pounds over the mark," Deal recalled.[59] Certainly most cattlemen would like to have such a problem.

With their crossbreeding program, the Greenbriar Ranch calving rate has moved from 55 percent to 85 percent, with mature crossbred cows getting 95 percent. Consistent with the Greenbriar's calving rate were studies made recently at the Louisiana Agricultural Experiment Station where 4-year-old F_1 Brahman-Angus cows calved 97 percent. Four-year-old Angus cows tested under the same conditions, had a calving rate of 80 percent. It was also demonstrated in these studies, that F_1 Brahman-Angus cows were higher in fertility rates at first and second service than were Angus or Brangus cows. Milk production was also much higher with F_1 Brahman-Angus heifers. For a 16-hour period tested over two years, milk yield for the F_1 heifers was 10.2 pounds while it was 7.5 pounds for the Brangus and 6.8 pounds for the Angus.[60]

Brahman-Crosses Continue to Win Support

"Brahman cattle have had a hard time in this state," said Alabama cattlemen Franklin Smart of Silverhill. "They just haven't taken here until recently. The people have always seemed a little afraid of the 'rodeo' image . . . ." But Smart remembers that when Brahmans were first crossed with woods cattle of Alabama, (descendants of early Spanish breeds), it added ". . . about 150 pounds to the average weaning weight . . . the brindle heifers tended to be real popular so most of the people tried to keep them and as a result they kept a lot ot Brahman crosses; because you see, they brindle easily."

Smart has raised Brahman cattle in Alabama since 1938, and concentrates his breeding program on the red Brahman. He gives a lot of credit to the Alabama Brahman Association for the great popularity of the Brahman in Alabama today.[61]

Elry Butler of Cheyenne, Oklahoma, also remembers earlier how neighbors of his "raised the dickens" when some of his Brahman bulls got in with some their cows. But ". . . there wasn't much complaining when the calves sired by his Brahman bulls sold for a dime a pound more than any of the other calves."

A close neighbor of Butlers, Eddy Cape of Merricks Red Rock Ranch, pointed out to Joe Brockett recently that Brahman cattle had become a solid part of their working ranch operation. In

Courtesy Santa Gertrudis Breeders International

Santa Gertrudis-sired replacement heifer with calf on the King Ranch, Kingsville, Texas. An infusion of Brahman blood into other breeds has made it possible to wean heavier cows without creep feeding. And on adequate pasture, the calves continue to gain weight at a rapid rate after weaning.

1974, they first used 80 Brahman-Hereford F_1's which weaned 78 calves, coming off the cows at a better than 600 pound average.

Later, Cape turned in some quality Brahman bulls from the Santa Elena Ranch at Madisonville, Texas with first calf Hereford -Limousin cross heifers.

Cape said that he has been particularly pleased with the fertility rate, the milking ability and the tolerance to insects of his Brahman-crosses, but most importantly, "They make money."[62]

On the G-Bar-E Ranch at Okeechobee, Florida, owned by Victor and Charlotte Clarke of Coral Gables, ranch manager Groover Long is using Brahman blood not only to produce heavier market calves and better replacement females, but he has been particularly committed toward producing more uniform calves. As the energetic and affable Long gently quipped, "I know you sell by the pound and you can't tell the difference with the hide off, but we sell our calves with the hides on and uniformity of type and color brings a premium." The G-Bar-E program calls for purebred Brahman and Angus. Utilizing the crossbred cows available, Long has grouped the cattle so that all straightbred Brahman cows are bred to Angus bulls and straightbred English type cows are bred to Brahman bulls. Long hopes to eventually utilize two families of linebred Angus bulls and linebred Brahman bulls exclusively to get maximum hybrid vigor and uniformity.[63]

Long keeps a purebred Brahman herd to produce good bulls for his program. When selecting new bulls, he looks for ample size, bone and muscling. "He selects alert, sound, masculine bulls with good performance behind them." Since he started his new program, weaning weights have increased on the average of 150 pounds. While he admits that outcross purebred bulls are sometimes structurally larger, he prefers the linebred bulls because of extra hybrid vigor and greater uniformity.[64]

. . . A Matter of Seed and Feed

One of the oldest issues among scientists has been the question of heredity versus environment. And the debate still continues - "Which is more important?" Cattlemen have debated the question just as vigorously. In speaking of livestock and crops, Thomas Jefferson was once reported to have said, "It's all a matter of seed and feed." While no enlightened cattlemen would challenge Jefferson's general thinking, there is still no consensus on which is the more important, seed or feed? - genes or range?

It would appear, as Jefferson implies, they are equally important, or at least, the difference is indeterminable over a broad range of circumstances. This was clearly illustrated when some cattlemen and researchers felt for a long time that the high rate of heterosis resulting from crossing Brahman and British breeds along the Gulf Coast region could only occur in subtropical or tropical regions. And that the culmination of this crossbreeding would not be nearly so evident in colder or more temperate environments. The fallacy of this thinking has since been demonstrated by breeders who have had outstanding success with Brahman crosses over rangeland regions where winters are consistently long and cold.

One of the most celebrated studies on cattle heterosis has been conducted during the past few years on some of the best, and coldest rangeland in North America, at Clay Center, Nebraska. In these aforementioned studies, the F_1 Brahman-hybrid females, in the Germ Plasm Evaluation Study, weaned 19 percent more pounds of beef than the Hereford-Angus dams, thought to be more suited to this range. The F_1 Brahman-hybrids also weaned 10 percent more calves.[65]

An earlier study carried out in Southern Alberta, Canada demonstrated that the longevity trait found in Brahmans was transmitted when American Brahmans were crossed with British cattle. Nineteen 2,3 and 4-year-old F_1 Brahman-crosses were compared with nineteen purebred British breeds of the same ages. After five years, there was a steady decline of performance and survival ability among the British breeds while the F_1 Brahman crosses maintained their inherited vigor. At the end of 14 years, all of the British purebreds were dead while three F_1 Brahmans were still alive - two were even pregnant.[66]

Cattleman Mike Mendenhall of Rock Ford, Colorado recently commented, "The F_1 Brahman cows will winter just as well as the Herefords and that is saying a lot, considering temperatures can reach 20° below zero in this country . . . In an average year calves out of our Brahman-cross cows will average over 50 pounds more than calves out of our Herefords, and in this business the bottom line is pounds produced per cow. The F_1 also tends to utilize pastures better, is more disease resistant . . . will calve as a two-year-old with very little difficulty and as a mother is superior to the straightbred Hereford."[67]

Unlike the Mule

"Notwithstanding its own beefiness, the Brahman is an indispensable ingredient of hybrid beef." Indeed, as Harry Gayden recently pointed out, "The most important reason for its being, for its value to the world beef industry lies in the hybrid vigor it brings through crossbreeding."[68]

This, of course, poses no threat to the purebred market. In fact, sales of purebred Brahmans have greatly increased through the years as better planned hybrid programs have been established. Commenting on the crossbreeding boom that was really getting off the ground in 1951, Brahman breeder Henry O. Partin of Florida said, "Crossbreeding has become popular is attested to by the fact that an estimated 90 to 95 percent of all purebred Brahman bulls are sold by breeders for use in commercial herds."[69]

Reflecting on recent changes in the cattle business, Brahman breeder Gilbert Tucker of Cocoa, Florida, whose family has been raising cattle in Florida since the 1850's, recently said, ". . . Now you see more and more Brahman-cross cattle. The more of them you see, the more Brahman bulls you are going to see follow. Of course, the Association's F_1 Certification Program will help considerably."[70]

But one must also consider the purebred Brahman female today. Earlier, virtually all hybrid cattle involving Brahmans were produced by the use of purebred Brahman bulls on cows of other breeds as Brahman cows were not readily available for commercial production. However, today high quality purebred cows are plentiful, and many hybrid breeders are employing the reciprocal cross to obtain maximum hybrid vigor.[71]

Even today though, the most feasible means of getting into hybrid production is by using the best obtainable Brahman bulls on the best cows already at hand. But hybrid vigor is "not a magical formula for creating new, superior genes from genetic trash. Superior hybrids . . . are produced by superior performing parental stock of a good type."[72]

Courtesy Adams Ranch, Fort Pierce, Florida

Although Brahmans have been crossed with Herefords for many years, the Braford breed was not actually established as a separate breed until recently. Above is a purebred Braford herd on a ranch in southeast Florida.

Unlike the complicated procedures of hybrid plant, swine and poultry production, outstanding hybrid beef cattle can be produced by simply crossing proven Brahman with proved European cattle, and unlike the mule, this F_1 progeny will be highly fertile and have the ability to transmit added vigor. Some researchers figure this advantage for crossbreds to be 5 to 10 percent, depending on the particular trait and how different genetically the two animals are that are crosses. When the first cross, the F_1 Brahman, is utilized as a dam, this can boost the production advantage to 18-25 percent over the average of the straightbred parents.[73]

"No amount of scientific skill and money has been spared by discerning American cattlemen in the developement of the magnificient . . . beef type Brahman of the Americas from the Indian Zebu." And the sturdy, hybrid progeny of the Brahman have practically girthed the globe "to tinge the blood of more than 50 percent of the cattle of the world."[74]

Just as the blood of the early Arabian horse has brought a special quality to many other breeds of horses, "the primal vigor of the Brahman"[75] has provided the general tenacity and the longevity for many new breeds. It will most assuredly be used for the creation of still new breeds for other environments - environments, perhaps, where cattle cannot be commercially produced today. Yes, the American Brahman is certainly "crossbreeding's common denominator."

In other lands they live and breed
to linger languidly at feed,
to render antics tinged with mirth
and do their bit for hungry earth.

Elizabeth Stilwell

8

ABBA Continues to Serve

We have seen how the founding of the ABBA in 1924 and its continued - oft times - dramatic growth helped to establish wide acceptance for the American Brahman as a superior beef animal. While it is, of course, the adaptability, vigor and versatility of the animal itself that has made the American Brahman so popular among so many breeders, the Association has provided a realistic, yet demanding, criterion for registration. It has been this model of excellence required by the ABBA that has caused cattlemen seeking the best Brahmans to insist upon ABBA papers.

The importance of having such a reputable association for the recordation of a particular breed can hardly be overstated; however, the ABBA has also provided a number of other important services over the years. Probably no single breed association in history has done more to advance its breed or to inform the public of its breed's unique qualities.[1]

Advertising

Advertising through various trade journals and other media has become an important *modus operandi* of the Association's

ABBA Director Joe Dial of Placedo, Texas, like numerous other ABBA Directors, has travelled to many parts of the world to serve on inspection teams, to participate in Brahman seminars and to promote markets for the American Brahman. Joe Dial and his wife Emily own the 3X Ranch at Placedo, Texas.

educational and promotional work. While its small membership during the first years struggled to initiate a registration program while also dealing with the pic-n-shovel duties of establishing a new organization, they had the early foresight to recognize and appreciate that any sustaining, viable association would need to promote the product of its membership.

Although the Association had no official breed publication during those early years, by the mid-1940's it was advertising in the *Brahman-Breeder Feeder, The Coastal Cattleman, The Cattleman, The Florida Cattleman, Texas Livestock Journal, The Ranchman*, and the *Western Livestock Journal.*[2] In July of 1947 the Board of Directors authorized the ABBA Publicity Committee to spend up to $2,000 for that year, a modest sum even for the time. By 1973, the Association was designating $8,000 per annum for advertising while using sixteen different publications for a schedule of 38 full page ads. The advertising budget was raised to $40,000 per annum in 1978 and was that same amount in 1979 and 1980.[3]

Although it is often difficult to directly relate advertising to increased sales, in 1974 the Brahman breed had its best twelve months ever regarding total sales and prices received. Ironically, it was a year as ABBA Director J. A. Clifford described

". . .that reaped the effects of government controls, skyrocketing interest rates, shortages, inflation, political upheaval and the biggest slide in commercial cattle prices our industry has ever seen." Despite such difficult times, in 1974, treaty sales showed record dollar volumes with 14 auction sales grossing 3.75 million dollars. And of 1,853 registered Brahmans transferred, they brought an impressive average price of $2,022 per head.[4]

During this same period, nearly 3,300 registered American Brahmans were also transferred to breeders and other interests outside of the United States. This impressive growth in Brahman export sales appeared in part to be the results of contacts made by ABBA and FAS (Foreign Agricultural Service) teams visiting outside of the United States. Importantly, also was the fact that foreign buyers were beginning to recognize and appreciate the qualities of the American Brahman.[5]

The Brahman Journal, recognized for some years by the ABBA as the official breed publication for Brahman cattle, is circulated all over the world with over 1,500 copies mailed each month to Brahman breeders and interests in foreign countries. Sixty-five copies go to U.S. Agricultural attaches in foreign countries located in the Southern Hemisphere. There are also some 5,000 domestic subscribers. This fine publication is noted for its well designed layouts, its informative articles, its news coverage and its high quality color ads.

Over the years *The Brahman Journal* has greatly helped to promote the Brahman breed as well as providing a professional vehicle for communication.[6]

The Association's Public Relations Committee, which eventually became a part of the present Communication Committee, has also effectively maintained a steady flow of correspondence with people all around the world interested in the American Brahman. In 1972, for instance, over 8,000 separate inquiries were received by the ABBA's main office. Each inquiry was religiously answered, and in most cases, additional educational brochures were mailed out.[7]

Since 1968, one of the most effective educational and promotional tools underwritten by the Association has been a color film dealing with the importance of the Brahman, appropriately entitled *Impact.* Distributed by Modern Talking Pictures and U.S. foreign agricultural attaches around the world, this film explains the tremendous influence that Brahman-blood has had toward improving beef and dairy breeding throughout the world. (See Chapter IX, "A Breed for All Seasons.") It has been viewed literally by millions of people in all four corners of the globe, and it continues to be a favorite with many agricultural interests.[8]

Since 1948, the ABBA has also employed a full-time executive-secretary* who has been in charge of different promotional activities regarding the Brahman breed and Brahman crosses. Part of the duties of this office have included the compiling and printing of circulars and other literature on Brahman cattle, promotional activities related to livestock shows and youth groups, such as the FFA and 4-H Clubs, and the encouragement and coordination of research by the universities and other institutions involving Brahman-blooded cattle.[9]

Premium Money

Since 1939, the Association has consistently helped to underwrite the cost of prizes and premiums for reputable livestock shows in which Brahmans have competed. During that year, the fledgling association authorized $500 for prizes to be awarded at the Houston Fat Stock Show and Livestock Exposition for winning Brahman entries. During the War Years (1941-1945) much of the money that would have normally gone for premium money was used for the purchase of U.S. War Bonds. But by 1947, the ABBA Directors authorized the contribution of over $4,000 for Texas state cattle shows alone. That same year, $1,500 was also authorized by the ABBA Directors for the Ocala

*This office is now entitled the Executive Vice-President (ABBA Minutes, February, 1981).

The late, revered "Slim" Partin of Kissimmee, Florida is seen here standing behind the premium winning *Lady C Emperor,* PH 920, ABBA 44967. This *Emperor* sired female was the Grand Champion on the Florida circuit during 1948 and 1949 winning at shows in Tampa, Ocala, Kissimmee and Webster.

Brahman Show and $1,250 for the Louisiana State Brahman Show. All in all, some $10,000 was appropriated during 1947 for prizes and awards presented at shows where purebred Brahmans were competing.[10]

As the number of shows in which Brahmans competed continued to rise, along with the cost of awards, the ABBA also continued to raise the amount of money budgeted for premiums. By 1976, $20,000 was allocated for award monies, but in 1980, this amount had been raised to over $45,000. Part of the money now awarded goes for premiums at carcass contests.[11]

Much of the work today regarding ABBA endorsed shows, such as the selection of judges, the recommended amount of money for premiums and the sites for some shows is decided upon by the Association's Show Committee.

Register of Renown

Besides its established tradition of helping to share the cost of reputable Brahman shows, the ABBA has also taken other measures to recognize and spotlight superior animals within the registry. One such method has been the founding of the "Register of Renown"-started in 1964. This has been sort of a hall of fame for Brahman sires and dams having outstanding progeny. According to Brahman breeder Dean Blackwell, "The Register was established. . . to identify and recognize sires and dams of progeny with outstanding show ring performance in association with approved shows."[12]*

In order for a Brahman bull to enter the Register of Renown, it was established by the ABBA that he accumulate at least 150 or more points earned by five or more of his progeny in the show ring. Female Brahmans, on the other hand, must have had 40 points or more earned by two or more progeny. As mentioned earlier, *Sugarland's Rexcrata #1* headed the 1979 cumulative register for bulls with a total of 1,451 points earned by 44 progeny. A 1971 entry, *Miss Glen Sup Manso*, now dead, heads the female register with 246 points (1979), accumulated by three progeny.

Today there are 32 bulls and 85 females that have entered the Register of Renown.[13]

*Leon Locke, former ABBA President, is normally given credit for the Register of Renown program. Through his commitment to the idea it was eventually endorsed by the ABBA Board of Directors.

Register of Renown, 1981

Bulls — Sires with 150 points by five or more progeny.

*Dead Males

Rank	Name	Ph. No.	Reg. Number	Calved	Entry Year	No. of Progeny Earning Points			No. of Points	Breeder	Principal Exhibitor
						Bulls	Females	Total			
1	Sugarland's Rexcrata	229/3	160018*	1963	1971	23	21	44	1451	U.S. Sugar Corp.	3W Ranches
2	RQ's Rexcrata 455	455	234287	1972	1978	12	18	30	643	M. M. Whitaker	Diamond T Ranch
3	JDH Loxey Gamel de Manso	55	154295*	1962	1971	21	15	36	606	Forgason-Koonce Williams	U.S. Sugar Corp.
4	Mr. 3X Mucho Grande	34/8	195356	1968	1977	7	5	12	436	Mr. & Mrs. Joe Dial	Mr. & Mrs. Joe Dial
5	Sugarland's Esto 53	11/5	208491	1970	1979	8	8	16	411	U.S. Sugar Corp.	V8 Ranch & Santa Isabel
6	CMI Rafles	101/7	259438	1967	1979	16	9	25	403	Clemente-Maitret Irisson	Dr. Gonzalez
7	Mr. V8 763/1	763/1	203763	1970	1979	7	9	16	373	V8 Ranch	V8 Ranch
8	Sugarland's Loxcrata 98	561/5	221863	1971	1979	2	6	8	349	U.S. Sugar Corp.	V8 Ranch
9	WHS Esto Imperator	479/1	127744*	1958	1968	10	9	19	348	W. H. Stuart	Seaberg Ranch
10	JDH Ellery M Manso	363	170757*	1965	1971	3	7	10	347	J. D. Hudgins Magnum Div.	J. D. Hudgins
11	MEW Crepusculo	05/2	224173	1971	1980	3	5	8	344	M. E. Wentz, Jr	M. E. Wentz, Jr.
12	JDH Maru de Manso	347	175848	1966	1973	11	3	14	320	J. D. Hudgins Hudgins Div.	J. D. Hudgins, Inc.
13	BBB 61	61	213730	1971	1978	4	10	14	316	E. O. Doggett	J. P. Wood, MD
14	JCT Tony T Manso	332/1	173587	1966	1974	7	9	16	286	J. Carter Thomas	J. Carter Thomas
15	JMD Elefante Manso	226/3	200701	1968	1978	7	5	12	267	J. M. Dellinger	Seaberg Ranch
16	WHS Ben Imperator 143/5	143/5	180410	1966	1975	8	2	10	259	W. H. Stuart	W. H. Stuart Ranch
17	Lok-Mor Suville 1	3/161	157390*	1963	1975	12	6	18	243	Lock Moore & Co. Ltd.	U.S. Sugar Corp.
18	Rio Cherokee 978	978	161522*	1963	1977	4	2	6	239	C. E. Yoakam	Henry Koontz
19	JTW Didor Ferno Manso 449/1	449/1	227236	1972	1980	9	1	10	232	J. T. White	J. H. Jefcoat
20	Mr. 3X Mucho Mucho	246/3	245631	1973	1980	3	3	6	229	Mr. & Mrs. Joe Dial	Mr. & Mrs. Joe Dial
21	Cherokee Negro 1/609	1/609	205930	1969	1977	8	3	11	225	C. E. Yoakam Estate	Cherokee Ranch
22	JDH Crat Moso Manso	982/7	115291*	1956	1968	7	3	10	217	J. D. Hudgins	J. D. Hudgins, Inc.
23	Ward's 2/34	2/34	238097	1972	1979	7	6	13	216	A. W. Ward	J. Williams & Flying V Ranch
24	IW's 361	361	221699	1971	1980	5	5	10	212	I. W. Whitaker & Sons	David Wilson
25	JDH Comas Manos	773/8	139812*	1960	1971	12	10	22	211	J. D. Hudgins	Runnells-Pierce Ranch
26	Sugarland's Albacrata Jr. 74	557	90044*	1953	1968	7	4	11	201	U.S. Sugar Corp.	U.S. Sugar Corp.
27T	JK Sugar Crata 45	45	256950	1974	1981	12	8	20	200	Jack Keel	Jack Keel
27T	CCPR Iman 10/4	10/4	280389	1974	1981	3	6	9	200	Cristobal De Castro	James P. Wood Ranch
28	HG Vernon 42/3	42/3	245603	1973	1981	2	4	6	193	Henry Koontz	Henry Koontz
29	MN Imper Manso 448	448	154114*	1963	1971	3	5	8	184	Bar M Ranch	Bar M Ranch
30	JDH Bano Cutler Manso	2	150110*	1962	1971	7	6	13	180	J. D. Hudgins	J. Carter Thomas
31	JDH Selim Rex H Manso	261	168913*	1965	1976	4	1	5	178	J. D. Hudgins	J. D. Hudgins
32	IW'S Rexcrata 402	402	221717	1971	1979	6	2	8	177	I. W. Whitaker & Sons	Francisco Davila
33	IW'S Rexcrata 262	262	199817	1969	1979	3	8	11	161	I. W. Whitaker & Sons	V8 Ranch
34	AJM Imperator Marso 954/3	954/3	165105	1964	1974	5	2	7	158	A. J. Marceaux	Russell R. Greene
35	IW's Sugar Rexcrata 201	201	191528	1968	1981	5	10	15	156	I. W. Whitaker & Sons	Wendell M. Johnson

Females — Dams with 40 points by two or more progeny.

*Dead Females

Rank	Name	Ph. No.	Reg. Number	Calved	Entry Year	No. of Progeny Earning Points: Bulls	Females	Total	No. of Points	Breeder	Principal Exhibitor
1	Miss Beth 1/3	1/3	279077	1972	1978	4	4	8	306	M. E. Wentz	M. E. Wentz
2	Miss Glen Sup Manso	92/7	140244*	1956	1971	1	2	3	246	Hudgins Bros.	J. D. Hudgins, Inc.
3	JDH Miss Didor Manso 13	413	216071*	1966	1972	2	1	3	194	J. D. Hudgins Mangum Div.	Neal Barfield Ranch
4	Whit's 251	251	240861	1969	1976	5	1	6	181	I. W. Whitaker & Sons	Jack Keel
5	Miss 3X King Grand 47	47/0	256377	1970	1978	0	3	3	176	Mr. & Mrs. Joe Dial	Mr. & Mrs. Joe Dial
6	Miss 3X Mucho Grande 133	133/2	267576	1972	1978	0	2	2	164	Mr. & Mrs. Joe Dial	Mr. & Mrs. Joe Dial
7	SW Saddle's Fancy Lady	300	330157	1976	1981	0	2	2	159	Sloan Williams	Sloan Williams
8	Double I's Queen de Manso 113	113	252172	1970	1980	2	0	2	156	Flying J Ranch	J. H. Jefcoat
9	GLP Miss Esto Galan 463	463	178066	1961	1968	2	3	5	153	G. L. Paret	H. L. Myers
10	JDH Lady Errol Manso 5	315	205301	1965	1972	2	0	2	151	J. D. Hudgins Magnum Div.	3W Ranches
11	Miss Faye 30/2	30/2	368279	1971	1980	1	1	2	144	M. E. Wentz, Jr.	M. E. Wentz, Jr.
12T	Buffs Country Girl 724	724	193323	1963	1973	2	0	2	140	Frank Buffaloe	3W Ranches
12T	HCK Princess Val Suvan	192/1	277193	1972	1980	2	0	2	140	Koontz Ranch	Watts Ranch
13	JDH Lady Rous Manso 729	729	272859	1972	1980	1	1	2	139	Hudgins & Sons	R. Richard
14	Miss 46	46	205180	1965	1976	2	3	5	137	I. A. Carpenter	D. Wilson 4C Ranch
15	Miss HK Vernon Arauto 927	927	243488	1969	1977	0	3	3	136	H. C. Koontz	H. C. Koontz
16	MN Miss Mayromanso 217/1	217/1	263694	1971	1978	1	1	2	129	Bar M Ranch	J. P. Wood, MD
17	Miss 3X V8 Tippu	25/8	234652	1968	1978	1	1	2	123	Mr. & Mrs. Joe Dial	Mr. & Mrs. Joe Dial
18	Miss HG 177	177	280569	1972	1980	1	1	2	120	Hales Farms, Inc.	Diamond T Ranch
19T	Miss Sahaswan Shanghai	439	198194	1964	1971	2	2	4	119	Runnells-Pierce Ranch	Runnells-Pierce Ranch
19T	ABF Lady B Chunko Manso 442	442	191868	1963	1972	3	2	5	119	Albert B. Fay	I. W. Whitaker & Sons
19T	JCT Miss Tommy Manso 780	780	170856*	1959	1973	1	1	2	119	J. Carter Thomas	J. Carter Thomas
19T	JCT Miss Holly T Jumbo	532/1	226787	1967	1974	3	2	5	119	J. Carter Thomas	J. Carter Thomas
19T	ABF Lady de Manso 123/1	123/1	274651	1972	1979	1	2	3	119	Albert B. Fay	Albert B. Fay
20	ADS Queen de Manso 708/1	708/1	175796*	1961	1975	1	1	2	114	A. Duda & Sons	A. Duda & Sons
21	Miss V8 218/1	218/1	203218	1965	1978	3	2	5	113	V8 Ranch	V8 Ranch
22	Miss Het of Sugarland 25	736/2	181886*	1961	1970	2	2	4	111	U.S. Sugar Corp.	U.S. Sugar Corp.
23T	JCT Miss Equero Manso 117/1	117/1	191205	1963	1968	4	0	4	109	J. Carter Thomas	J. Carter Thomas
23T	JDH Miss Stando Manso 44	444	216096*	1966	1972	0	4	4	109	J. D. Hudgins Magnum Div.	J. D. Hudgins Magnum Div.
24	Miss V8 887	887	178887	1961	1980	2	0	2	100	V8 Ranch	V8 Ranch
25	AJM Miss Topper Marso 283	283/3	189916	1963	1978	1	1	2	106	A. J. Marceaux	Russell Roy Green
26	Miss V8 560th	560	144560*	1956	1972	1	2	3	98	V8 Ranch	H. D. Pruett
27T	Rio Queen 696/1	696/1	234470	1968	1978	1	1	2	95	Clifford Bros.	Dean & Neil Blackwell
27T	Miss Cherokee Arauto 1/972	1/972	277217	1972	1978	2	0	2	95	Cherokee Ranch	L. Moreau & Long's Ranch
28	JDH Lady Didor Manso 59	159	192533	1963	1973	2	0	2	93	Ted Mangum	J. D. Hudgins Inc.
29T	L6 Christina Manso 528	528	176693	1961	1971	0	3	3	92	L6 Ranch	L6 Ranch
29T	Miss V8 146/1	146/1	196146	1964	1976	0	2	2	92	V8 Ranch	V8 Ranch
29T	JDH Lady Oka Manso 135	135/3	261619	1971	1980	2	1	3	92	J. D. Hudgins Forgason-Koonce	C. Hudgins
30	5P Miss 226	226	299080	1974	1981	0	2	2	89	Powell Farms	J. P. Wood
31T	MN Miss Jenny Kee Manso 471	471	188356	1963	1970	2	1	3	88	Bar M Ranch	Bar M Ranch
31T	MCT Miss Tomtop Manso 187/1	187/1	197280	1964	1976	1	1	2	88	J. Carter Thomas	J. Carter Thomas
32	HFB's 639	639	193293	1963	1972	0	2	2	86	Hugh Buffaloe	3W Ranches
33	ABF Lady Houston	446	191872	1963	1971	1	2	3	85	Albert Fay	3W Ranches

34	Miss BBB 153	153	278431	1972	1981	1	2	3	82	BBB Ranch	J. P. Wood, MD
35	Double J's Mixx Rex Imperator 2	117	252464	1970	1980	2	0	2	81	J. H. Jefcoat	J. H. Jefcoat
36	JDH Lady Katie Manso	222	196453	1964	1976	0	2	2	79	J. D. Hudgins Hudgins Div.	J. D. Hudgins Inc.
37	JDH Lady Rha Manso 19	719	305847	1965	1978	1	1	2	76	J. D. Hudgins	J. D. Hudgins
38T	ADS Queen de Manso 676/0	676/0	139567*	1956	1966	0	2	2	75	A. Duda & Sons	A. Duda & Sons
38T	JCT Gorgi T Shanghai	282/2	284876	1973	1981	2	0	2	75	J. Carter Thomas	J. Carter Thomas
39	JDH Miss Phila Manso	864/2	256819	1970	1978	2	1	3	74	J. D. Hudgins	Francisco Solano
40	Miss Sugarata 36	36	271678	1972	1979	2	0	2	72	R. Q. McSwane	J. H. Novak & Sons
41T	Queen of Sugarland 6	13/2	164483*	1959	1968	2	0	2	71	U.S. Sugar Corp.	U.S. Sugar Corp.
41T	Miss C of Sugarland 33	686/2	181862*	1961	1974	5	0	5	71	U.S. Sugar Corp.	U.S. Sugar Corp.
41T	HCK Lady Suva V Nobell	194/1	277194	1972	1979	0	3	3	71	Koontz Ranch	Watts Ranch
41T	Miss Mc Cherokee Rio Negro 329	329	221618	1967	1980	0	2	2	71	J. M. McKellar Estate	Hudson Cattle Corp.
42	JDH Lady Wilery Manso 306	306/8	174773	1960	1967	2	0	2	70	J. D. Hudgins	J. D. Hudgins, Inc.
43T	Queen of Sugarland 4	962/1	164460*	1958	1972	2	2	4	68	U.S. Sugar Corp.	U.S. Sugar Corp.
43T	JDH Lady Coro de Manso 501	501	241889	1969	1979	1	2	3	68	J. D. Hudgins Locke Div.	Leslie W. Hudgins
43T	JDH Lady Orestes de Manso 0/3	930/3	282290	1973	1981	1	1	2	68	J. D. Hudgins Forgason-Koonce	J. D. Hudgins Forgason Div.
44	WHS Cornelia Imperator 699/1	699/1	165050	1959	1976	2	0	2	63	W. H. Stuart	W. H. Stuart Ranch
45	Miss Resoto of LSU 20	735	228569	1968	1980	3	0	3	62	LSU—Baton Rouge	Berchman Laverge
46	Miss AP 139	139	191070	1963	1971	3	0	3	61	AP Ranch	AP Ranch
47T	Miss AP 350	350	222118	1967	1977	1	1	2	58	AP Ranch	AP Ranch
47T	Miss 3X Ipteb 205	205/3	287242	1973	1980	1	1	2	58	Mr. & Mrs. Joe Dial	Mr. & Mrs. Joe Dial
48	Miss Paret Ranch 108	108	130927*	1955	1969	1	1	2	56	G. I. Paret	G. L. Paret &
49T	Miss R-Pierce Ranch 308	308	182659	1962	1973	1	2	3	55	Runnells-Pierce Ranch	Runnells-Pierce Ranch
49T	Miss 50	50	205456	1965	1980	1	1	2	55	I. A. Carpenter	David Wilson
50T	Miss Manso 112	112	198705	1964	1975	1	2	3	53	M. M. Whitaker	Seaberg Ranch
50T	Miss V8 707/1	707/1	244707	1970	1980	1	1	2	53	V8 Ranch	V8 Ranch & C. Thomas
51T	Miss Carolyn Metz	625	256111	1971	1977	1	2	3	52	G. A. Tucker	WW. or Martha Story
51T	Miss V8 866/1	866/1	254866	1971	1979	1	1	2	52	V8 Ranch	V8 Ranch
51T	JDH Lady Quillo Manso 771	771	261665	1971	1980	1	1	2	52	J. d. Hudgins Hudgins Div.	J. D. Hudgins Hudgins Div.
52T	Miss Paret Ranch 152	152	139039*	1956	1969	0	2	2	51	G. L. Paret	G. L. Paret & Sons
52T	Lady Arnet de Manso	506	155487*	1958	1974	1	2	3	51	S. C. Border	J. D. Hudgins, Inc.
52T	JJ's Miss Esto Imeprator 140	140	270294	1971	1980	1	1	2	51	J. H. Jefcoat	J. H. Jefcoat
53T	Miss V8 539/1	539/1	226539	1968	1979	1	1	2	49	V8 Ranch	V8 Ranch
53T	Miss V8 194/1	194/1	203194	1965	1979	1	1	2	49	V8 Ranch	V8 Ranch
54	JDH 971 Jake Manso	971/6	132641*	1955	1965	1	2	3	47	J. D. Hudgins	J. D. Hudgins
55	Miss 3XAA Shot Guns Brazindu	16/5/AA	208180	1965	1980	2	0	2	46	Mr. & Mrs. Joe Dial	Mr. & Mrs. Joe Dial
56T	Miss Het of Sugarland 34	440/3	201713	1964	1972	0	2	2	45	U.S. Sugar Corp.	U.S. Sugar Corp.
56T	GLP Miss Rex Rodrigo	784	214191	1966	1974	2	0	2	45	G. L. Paret	Richard & Robert Smith
56T	Miss V8 615	615	156615	1958	1979	0	2	2	45	V8 Ranch	V8 Ranch
57	Miss JMD 823	823	203717	1964	1973	2	0	2	44	J. M. Dellinger	I. W. Whitaker & Sons
58T	Miss JH's 138/2	138/2	266639	1970	1977	0	2	2	42	J. H. Brazier	Courtney Stock Farm
58T	58 OLP Tonto Bano Manso	847/2	270206	1972	1979	2	0	2	42	O. L. Partin	O. L. Partin
58T	Miss Kikkroy Imp 14	14	199321	1964	1980	1	3	4	42	I. W. Whitaker & Sons	International Ranch Inc.
59T	SW Lady Cherokee 80	80	247682	1970	1977	2	0	2	41	Sloan Williams	Mr. & Mrs. Joe Dial
59T	RPR Miss Adam 801	801	244473	1969	1979	2	0	2	41	Runnells-Pierce Ranch	Runnells-Pierce Ranch
59T	HCK Miss Corrine Flato	780	188598	1963	1981	2	0	2	41	Koontz Ranch	Jim D. Brown
60T	WMJ Miss Resoto Manso 16	16	217366*	1966	1974	0	2	2	40	Wendell M. Johnson	Wendell M. Johnson
60T	GLP Miss Esto Tippu 449	449	178052	1961	1970	1	1	2	40	G. L. Paret	G. L. Paret & Sons

One of the outstanding bulls listed on the ABBA Register of Renown is *Mr. V8 763/1.* He was bred and raised on the V8 Ranch at Hungerford, Texas, owned and operated by Sloan, Mollie, Jim, Luann and Jan Williams. The V8 Ranch has received 17 Register of Renown plaques. and over the past five years has been the most consistent winner in the Brahman show ring. Developed from W.S. Jacobs and J. D. Hudgins bloodlines, the V8 herd is one that consistently breeds true to form. One of the most important herd sires in the V8 breeding was *JDH Rex Emperor de Manso,* PH 470/6. (Interestingly, the V8 herd received its name from Howard Parker of Center, Texas, who owned the first V8 herd. Parker, who also owned a Ford agency, liked the V8 engines so much that he chose the name for his herd.)

Photo above - ABBA Director Cecil K. Boyt of Devers, Texas (at left) served as an Association director longer than any other director in ABBA history. He is shown here in 1955 with ABBA President J. T. White of Hearne, Texas and former ABBA President Edgar Hudgins of Hungerford, Texas.

Photo right - J. T. "Jack" Garrett of Danbury, Texas is the second man to have served on the ABBA Board of Directors for over 30 years.

Officers and directors for the 1980 National Junior Brahman Association recently elected the following officers (top row from left): Anya Haidusek, Director; Tawyna Grooms, President; Tammie Watts, Secretary; Gerald Tice, Reporter. Bottom row from left, Lucas Hinojosa, Director; Stephen Kohara, Director; Chuck Lahaye, Treasurer; Chad Fontenot, Director and Russ Gould, Vice-President.

American Junior Brahman Association

In its long commitment towards youth oriented activities and youth organizations, the ABBA, under the dedicated steerage of Wendell Schronk and Lee Pritchard, has recently organized the American Junior Brahman Association. This new organization was established primarily to involve youth in the raising and showing of registered Brahman cattle. But it has also provided an organization by which young people can register and transfer qualified Brahmans through the ABBA. Today there are affiliate American Junior Brahman Associations in Texas, Louisiana, Arkansas, and Florida. While started only two years ago, the membership is already over 400.[14]

The most important annual event for the American Junior Brahman Association is its All-American National Junior Brahman Show and Meeting in which members are involved in show ring competition and other activities. Last year's (1980) show was held in Baton Rouge, Louisiana, where 280 registered Brahmans were entered by youth from ages eight to twenty-one. Judged by ABBA Guatemalan Director Neil Potter, the quality of the entries was unusually good. Brahman enthusiasts, exhibitors and breeders from many parts of the nation were in attendance.[15]

Today the ABBA maintains a permanent Youth Activities Committee established to help young Brahman breeders and to coordinate the activities of the AJBA with the other activities of the ABBA.

Research Programs

Since the 1940's, the ABBA and its individual members have taken an active role in helping defray the costs of worthwhile research projects involving Brahmans and Brahman-crosses. (See Chapter VI, "Brahmans Pass All Tests.") As we saw earlier, ABBA members have contributed to research programs for over 30 years with the outright donation or leasing of registered Brahmans. In 1972, the ABBA and some of its members helped provide approximately 500 animals through various arrangments for experimental work.[16]

For a number of years, the Association has also provided direct cash funds for research work. And in order to provide a more reliable funding for Brahman research, the ABBA established a regular annual research fund beginning in 1963. Since then, money for research has been included as a regular part of the ABBA's annual budget. From 1975 to 1980, approximately $70,000 was appropriated by the Association specifically for research. In order to evaluate research data and funding requests, the ABBA has maintained a Research and Breed Improvement Committee.[17]

ABBA Directors view Texas A&M University's herd at a board meeting on the campus at Col lege Station in November 1953. The herd was established in 1946 with a gift of six heifers fron the J.D. Hudgins Ranch, Hungerford, Texas. (Left to right) ?, A. G. Dingle, Wilbourn Gibbs, L. S Harris, ?, James C. Miller, Henry O. Partin, Bob Jones, Howard Parker, Ray W. Wilson, Vernor Frost, Cecil K. Boyt, Robert Schrode, Edgar Hudgins, ?, O. J. Flowers, Sidney L. Crochet, Bil Pate, G. A. Parr, Albert B. Fay, George Artecona, Bob Milam, A. N. Smith.

Texas A&M, Lousiana State University and the University of Florida have been particularly active in research involving Brahmans and Brahman-crossbreeding. Typical of the experimental projects currently underway, that is being partially underwritten by the ABBA, involves the blood typing of Brahman cattle. Headed by Dr. Jerry Caldwell of Texas A&M, one of the aims of this project has been to expedite and make more accurate the blood typing of Brahman cattle. Blood typing for Brahman cattle over the years has generally taken a long time because of a lack of information as to the various phenotypic groups of Brahman cattle. Caldwell's important research has already greatly improved the accuracy by which "two sire problems" in Brahman cattle can be determined.

"An important phase of Dr. Caldwell's work on blood typing," according to Wendell Schronk, "has been the study of white cells which give off antibodies thought to make Brahmans more disease resistant than other breeds."[18] In both 1980 and 1981 the Association appropriated $25,000 for this particular research program.[19]

Back in 1966, the ABBA started a study on the possibility of instituting a herd improvement program for Brahman breeders. This same program was eventually implemented and designated as the Brahman Herd Improvement Records Program. Originally its efforts to encourage breeders to constantly improve their Brahman herds came under the auspicies of the ABBA's Research Committee. Today this committee is now simply referred to as the Research and Breed Improvement Committee. Information made available by the Association through the BHIR records has provided an excellent way for improving Brahman and Brahman-crossbred herds.

By 1977 over 100 prominent breeders were participating in this program.[20]

With a continued commitment toward consumer education, in 1974 the Association's Directors passed a resolution in support of the Beef Development Taskforce Program. While it was completely voluntary, the program was voted down by the industry. The overall program was expected to generate some $40 million dollars for the promotion and research for all sectors of the cattle industry.[21]

The Promotion of Foreign Markets

During the late 1950's and the early 1960's, recorded foreign sales of American Brahmans began to show a dramatic increase. For instance, in 1959 nearly 5,000 registered Brahmans were recorded as foreign sales - nearly fifty percent of the total recorded sales for registered Brahmans for that year. In order to help maintain this interest by factors outside of the United States and to facilitate communication with breeders and other interests involved in international trade, the ABBA established its In-

The American Brahman Breeders Association and the FAS of the USDA have been highly successful in popularizing the American Brahman in many parts of the world. This herd scene of Brahmans and Brahman-crosses is in Northern Argentina.

ternational Committee. This particular committee has become one of the most active and productive of all the Association's standing committees.[22]

The International Committee has often worked jointly with the Foreign Agricultural Service of the USDA throughout the world in investigating and developing foreign markets and in expanding existing markets. In the belief that there is no substitute for personal contacts, the International Committee has helped to plan and finance foreign trips for ABBA staff and ABBA members travelling outside of the United States to meet foreign cattle breeders and to judge foreign Brahman shows.[23]

These joint ABBA and FAS teams have done an excellent job in keeping Brahman breeders and other interests informed about world market conditions. They have performed an invaluable service in making foreign interests more aware of the goals of the ABBA and in helping to advertise the services provided by the ABBA. This cooperative effort between the ABBA and the USDA's Foreign Agricultural Service began as a brainchild of ABBA President Ben Carpenter and ABBA Executive Secretary Harry Gayden. "This aggressive effort was almost solely responsible for ABBA's breakthrough into the 'big time' making Brahmans the leading export breed from the U.S. And through the years, these teams have provided an important way for the exchange of cattle management ideas. After a recent trip to Zimbabwe, Africa, ABBA's Wendell Schronk reported that cattle breeders in that country, emerging recently from one of Africa's longest revolutions, "were proud to see the U.S. interest in their country and cattle industry and profited from the management discussions conducted by our marketing team."[24]

In its continued commitment toward encouraging communication and competition between international Brahman breeders and in making the American Brahman still more visible around the world, the ABBA directors recently approved plans for an international Brahman show to be held in Houston, Texas in 1982. It was further proposed that national shows be continued on an annual basis with a different site each year and that ABBA sponsored national sales be held in conjunction with these shows.[25]

While the ABBA has historically sought to promote foreign interest in the American Brahman, wisely, it has been reluctant to accept every plan that might somehow make the breed more visible. In the '70's, for instance, a Brazilian based organization entitled COMZEBU, that wanted to establish a world wide Zebu association, invited the ABBA to become an affiliate member. Instead of becoming an "active" member of COMZEBU, ABBA Directors in 1978 voted to become only an "effective" member. ABBA Director Neil Potter seemed to reflect the feeling of most ABBA members when he told the COMZEBU's president that in ". . . his opinion until Brazil recognized the American Brahman as a breed, not only the United States, but all of the Central American countries that have the American Brahman type cattle should not join COMZEBU because Brazil is the headquarters of COMZEBU. . . ."[26]

Through the F_1 Female Certification Program, the ABBA has provided a means for recognizing the progeny from crossing two genetically unlike parents. Such progeny usually manifest the greatest hybrid vigor. These young F_1 and F_2 Brahman and European crosses are on a feedlot in Colorado.

The F_1 Female Certification Program

One of the most recent ABBA sponsored success stories has been the previously mentioned F_1 Female Certification Program. Conceived and promoted by ABBA's Executive Vice-President, Wendell Schronk, this program was implemented in 1979. One of the exciting and gainful results of this program has been the holding of several joint-field days involving ABBA affiliate associations with Hereford and Angus associations. Similar events are planned for 1981 which will involve ABBA affiliate associations with Beefmaster, Polled Hereford associations and possibly other breed associations.

While the F_1 Certification Program is less than two years old, over 1,000 F_1 heifers have been certified involving about 200 breeders from such states as Colorado, California, Arizona, Nevada, Oklahoma, Louisiana, Mississippi, Georgia, Texas and Florida. Of these breeders, the Pruett Wray Cattle Company of Sasabe, Arizona has the largest number of F_1 heifers registered as of 1980. Fred Jenkins of Ely, Nevada recently purchased over 200 registered Brahman bulls to use with his 2,200 Herefords in his F_1 breeding program. (See "Crossbreeding's Common Denominator.")

In his 1980 address to the annual ABBA meeting at Houston, ABBA President Lee Berwick described the F_1 cow as ". . . the best producing cow on the face of the earth."[27]

Growth and Expansion

The continued growth of the ABBA's services and responsibilities since its creation in 1924 has called for constant expansion of its facilities and the size of its staff.

During those formative years, specifically between 1924 and 1931, the Association was housed in a part of the old Livestock Exchange Building at 4905 Calhoun Road in Houston. It was during this time that the venerable J. W. Sartwelle served as the ABBA's Secretary.

ABBA HEADQUARTERS AT SIX LOCATIONS SINCE 1924

Livestock Exchange Building in Houston. ABBA's first office was in this building.

Leased second floor headquarters at 2711 South Main St. ABBA offices were on the second floor.

Leased headquarters at 1208 Louisiana. ABBA offices were on the second floor.

Association owned headquarters at 4815 Gulf Freeway.

New ABBA international headquarters at 1313 La Concha Lane.

On June 24, 1931, when Walter J. Hudgins was elected ABBA Secretary, the headquarters was moved to Hungerford, Texas where Hudgins lived. He served as Secretary until March 1, 1938, when Mrs. S. C. Border, his sister, began to serve as the ABBA's first official Recording Secretary. She very ably served in this capacity until 1948.[28]

In 1947, the Association moved its headquarters back to Houston to a rented office at 2711 South Main Street. "At the

Through the years the ABBA has managed to maintain an outstanding, dedicated staff. Pictured here is the present staff (1981). Seated (left to right) Lee Pritchard, Director of Association Services; Dinah Weil, Recording Secretary; Wendell Schronk, Executive Vice-President; Dwilda Daw, Director of Accounting; Ce Cowart, Director of Communications; Ann Green, Asst. Recording Secretary.

Standing (left to right) Carol Hottois, Cindy Hejl, Tricia Hernandez, Ann Zukywich, Jo Ann Thompson, Terry Robinson, Laura Clark, Charles Greenwood, Delann Sherrod, Armalinda San Miguel, Cheryl Morris, Sandra Greenwood.

time the Association's physical assets totaled a safe and one typewriter." (This, of course, did not include the registration records that the ABBA had accumulated.) In 1952, the headquarters were moved again to a yet larger rented space at 1208 Louisiana Street in downtown Houston. Later the Association owned a building at 4815 Gulf Freeway.[29]

During the fall of 1971, the ABBA purchased a building site in Shell Oil Company's attractive and spacious Plaza Del Oro - near the Houston Astrodome Complex.* And on February 5, 1973,

*John Lyle, U.S. Congressman and ABBA Director, was instrumental in helping ABBA obtain a choice site at a reasonable price.

Since 1924, the ABBA has registered over 750,000 Brahman cattle. Pictured are (left to right) Hal B. Allen of Houston, Texas registering the No. 750,000 Brahman animal with ABBA Recording Secretary Dinah Weil and Executive Vice-President Wendell Schronk. The Association now registers over 35,000 head of Brahman cattle a year. The Association began registering cattle in 1924 and took 49 years to register 500,000 head. It took Brahman breeders only the past eight years to record the 750,000th animal.

the Association began operations in its present headquarters at 1313 La Concha Lane. It was the realization of a dream for many ABBA members when this modern $390,000 building was first opened. Perhaps no members were more instrumental in bringing about its completion than Clive Runnells of Bay City, Texas and John Lyle of Houston, Texas.[30] By 1973 the full-time staff of the ABBA had grown to 16 employees.

The Association has continued to modernize its services and systems, and currently it is in the process of computerizing all of its records. This will provide a much more efficient technique of tracking information for breeders and utilizing existing space.[31]

Considering the history of the American Brahman and the important role that the Association has played in the establishment and the evolution of this breed, the ABBA will undoubtedly continue to play a strategic role in the years ahead - indeed, probably even more so than it has in the past. Despite the fact that the ABBA has vastly increased the scope of its activities over the years, its philosophy and objectives have remained essentially the same since its founding:

> *The purpose and object of the Association is to promote the breeding of Brahman cattle by all lawful and proper means, to keep proper records showing the pedigree and transfer of ownership of all Brahman cattle entered in the Herd Register of this Association, to assist in the arrangement of cattle exhibitions and shows, to aid in the advancement of scientific education concerning the breeding of Brahman cattle and to do such other things as may be in the interest of the Brahman breed of cattle and none of which shall be for profit.*[32]

With the development of important foreign markets, the revolutionary impact of media advertising, the creation of important Brahman derivative breeds, and the phenomenal strides made in techniques of animal research, it is interesting that this early statement of purpose has remained so applicable through the years. Those early visionaries who labored during the difficult and formative years of the ABBA provided an example of foresight and commitment and produced a structure by which their early dreams surely have been exceeded. To remain the vital and vigorous force that it is today, the ABBA must continue to aspire for yet greater goals which should include the still wider, international acceptance of the American Brahman as a beef and dairy animal. It must also continue to enlarge its current membership and to continue seeking better ways to serve its membership.

The American Brahman Breeders Association has joined the ranks of the computerized world. A Datapoint 4650 system has been installed at the office in Houston, Texas to serve the over 3,318 active members throughout the world. The system is currently working on the accounting and membership sections of the ABBA, and upon their successful completion will begin the work of housing the entire recording section into the new system.

Pictured with the new ABBA computer system is (left to right): Dinah Weil, ABBA Recording Secretary; Wendell Schronk, ABBA Executive Vice-President; and seated is Dwilda Daw, ABBA Director of Accounting.

Past Presidents of the American Brahman Breeders Association

A.M. McFaddin
Feb. 1924—June 1931

J.F. Hutchins
June 1931—Aug. 1935

F.N. Neuhaus
Aug. 1935—March 1940

James W. Sartwelle
March 1940—Feb. 1945

J.M. Frost, Jr.
Feb. 1945—Feb. 1947

J. Afton Burke
Feb. 1947—Feb. 1948

Past Presidents of the American Brahman Breeders Association

Gail Whitcomb
Feb. 1948—Feb. 1949

Herman Taylor
Feb. 1949—Feb. 1951

Henry O. Partin
Feb. 1951—Feb. 1953

Edgar H. Hudgins
Feb. 1953—Feb. 1955

J.T. White
Feb. 1955—Feb. 1958

L.S. Harris
Feb. 1958—Feb. 1960

Past Presidents of the American Brahman Breeders Association

Ben H. Carpenter
Feb. 1960—Feb. 1962

A.N. Smith
Feb. 1962—Feb. 1963

G.L. Paret
Feb. 1963—Feb. 1965

G.T. Stack
Feb. 1965—Feb. 1967

Leon Locke
Feb. 1967—Feb. 1969

Robert L. Massey
Feb. 1969—Feb. 1971

Past Presidents of the American Brahman Breeders Association

R.W. Mayronne, Jr.
Feb. 1971—Feb. 1973

Max Hammond
Feb. 1973—Feb. 1975

John Runnells
Feb. 1975—Feb. 1977

Dr. Richard Forgason
Feb. 1977—Feb. 1979

Lee Berwick
Feb. 1979—Feb. 1981

Frank D. Connell
Feb. 1981—

ABBA Executive Staff Officers

ASSOCIATION SECRETARIES

James W. Sartwelle
1924-1931

Walter J. Hudgins
1931-1938

Mrs. S. C. Border
1938-1948

ABBA EXECUTIVE SECRETARIES

Harry P. Gayden
1948-1974
Exec. Secretary Emeritus
1974-Present

Kirby Cunningham
1974-1976

Wendell E. Schronk
1976-1981
Exec. Vice President
1981-Present

ABBA RECORDING SECRETARIES

Margaret Watkins
1948-1974
Employed 1942-1974

Lu Anne Wright
1974-1976

Dinah Weil
1976-Present

When Harry Gayden (at right) retired as ABBA Executive Secretary in 1974, he passed the torch on to the able Kirby Cunningham. Cunningham served as ABBA's Executive Secretary until 1976. During Cunningham's tenure the Association continued to make great progress with all-time records being set in registrations, transfers (sales) and new memberships.

9

Brahmans Take to the Air

Since the end of World War II, great changes have taken place in the transportation of cattle and other livestock. While such traditional methods as the movement of cattle by trucks and ships still remain very important, the use of air transportation has become extremely important for the exportation of cattle. This has been particularly true in the case of American Brahmans as there has been such a universal demand for this breed. The American Brahman has been exported much more from the United States than any other cattle breed.

Before World War II, cattle bound for foreign destinations were normally transported by ships with the exception of cattle that were trucked or sometimes railed to the contiguous countries of Mexico and Canada. A number of shipping lines maintained specially equipped ships for such transfers. Today, the C. Clausen Steamship Company Ltd. specializes in transporting livestock all over the world. It annually transports approximately 1.8 million head of livestock from sheep to cattle.* Some of its freighters can carry up to 3,000 head of cattle.[1]

*Interestingly, the Clausen company is deeply involved in the transportation of live sheep to the many Moslem countries of the Middle East and Asia. The Moselms use thousands upon thousands of live sheep each year for sacrifices during religious ceremonies. Many of these sheep come from Australia (J. D. Sartwelle, former ABBA Director, interview, Houston: Oct. 30, 1981).

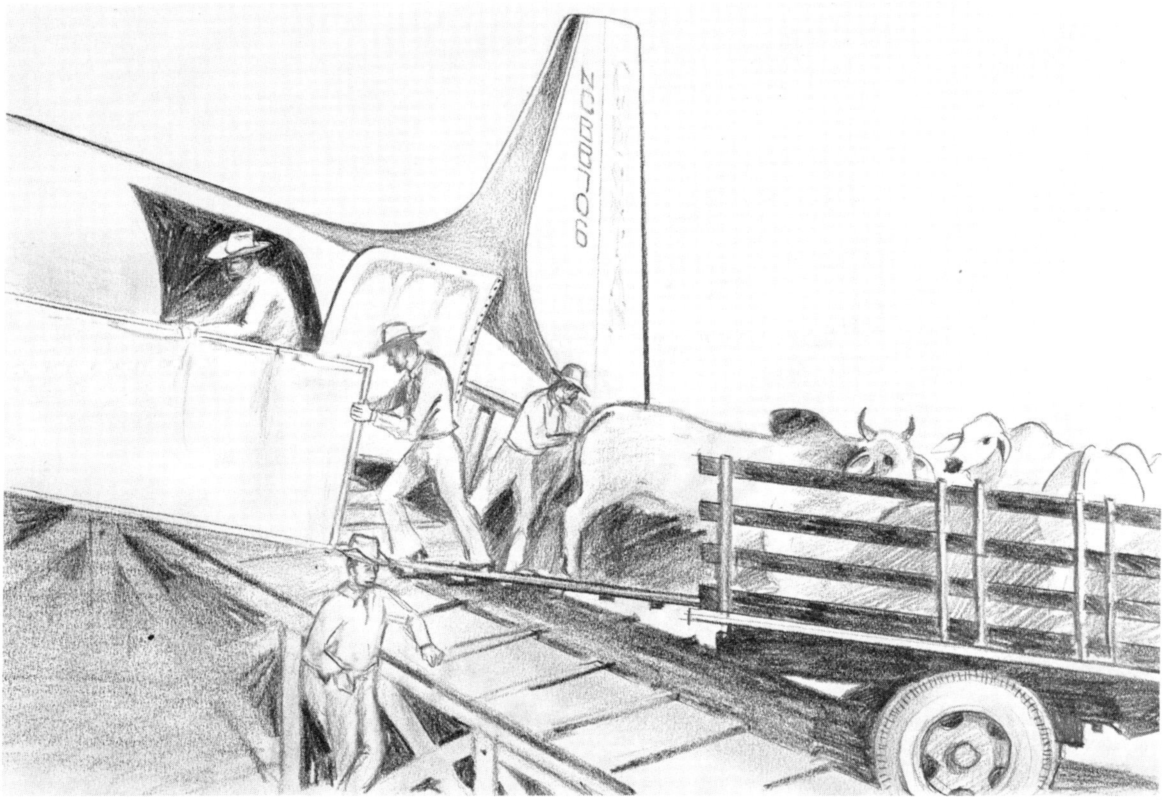

At the end of World War II, wartime workhorses, such as the C-47 and C-46 were soon put into service hauling Brahmans and other cattle to a hungry world.

When World War II ended in August of 1945, millions of people were left without adequate food, shelter or clothing. Many farms lay devastated, and these people generally looked to the United States and other affluent nations for help. Many private agencies as well as the U.N. quickly provided aid for the starving and homeless, but one of the immediate problems was one of distribution.

2

Millions of tons of produce, livestock and tools were transported by the sea lanes, but the war had dramatically shown how sizeable cargoes could be rapidly moved by air. The planes were small by today's standards, but they were still capable of moving large quantities of goods over long distances in short periods of time.

One of the first Brahman cattle operations to utilize airplanes for transporting cattle was the Norris Cattle Company of Ocala, Florida. Former ranch manager, Byron Johnson, recalls that when he started working at Norris in 1946 the company was already in the air transport business. "Cattle were loaded on planes right there in Ocala," Johnson recalls, "and shipped to Latin American countries. At first the Brahmans were loaded on C-47's, some of the real workhorses during the war, and the pilots were veterans who were used to hauling troops and military supplies rather than cattle. The payload was limited to about 6,000 pounds - a far cry from today's loads."[3]

During these early years of moving cattle by air, in Florida St. Petersburg became the terminal for transporting cattle overseas. There was a quarantine station there, and the facilities were excellent. In time, however, Miami became the largest air export terminal, and cattle were trucked in there from all around the country to be loaded on cargo planes. Eventually export facilities for the transfer of cattle were also established at Houston, Texas and other terminals.[4]

Another early pioneer in air-shipping Brahmans and other livestock was the Port City Stockyards - then of Houston - which had been in the livestock export business since 1937.*

J. D. Sartwelle, then Vice-president of Port City, remembers that when the war ended in 1945, few established airline carriers were interested in handling cattle. But in 1947 an air freight company known as Skytrain Airways offered to start carrying livestock. Made up of a group of gutsy, young World War II pilots, they were anxious to utilize the skills they had learned during the war.

During the early stages of Port City's experience in shipping cattle by air, there were a number of kinks that had to be worked out in loading and hauling these animals. C. R. Guthrie, former

*The Port City Stockyards are now located at Sealy, Texas; and its export division is operated by the state of Texas (J.W. Sartwelle, op. cit.)

stockyard superintendent, remembered that the first shipment of livestock to leave Houston by air was a mixed lot of Brahman bulls, mules and jacks destined for Latin America. "We had our minds set on crating . . . and I guess we weren't too smart about it at first," he recalled. "We would take a young bull weighing 1,000 to 1,400 pounds and box him up in a reinforced crate that weighed 650 pounds. Then we would take everybody and his dog along to the airport to help hoist that load onto the plane. Pretty soon though we got smarter. We put the crate on the plane and let the bull walk into it."[5]

Some of these early flights almost resulted in disaster. J. D. Sartwelle remembers the following incident that took place on an early flight:

The airborne animals had progressed beyond the individual crating stage by then, but they were tied, side by side, in the planes to keep the load from shifting.

On a run to Guatemala we were loaded to full capacity with mares, stallions, mules and cattle all mixed together . . . at the insistence of the Guatemalan representative who wanted some publicity shots of the animals being unloaded at Guatemala City.

The plane was way out over the Gulf when my worst fears were realized. A stallion and a jack had worked themselves loose enough to start a whale of a fight . . . the commotion set all the animals in a frenzy . . . they were raising so much sand that we couldn't get in among them . . . . The airline president, Ed McGee, was along on the trip and was dashing forward to get a pistol from the pilot when I reminded him that if we increased our altitude, the lack of oxygen would cause the animals to become docile. As the pilot leveled out at about 14,000 feet, the animals suddenly calmed down.[6]

J. D. Sartwelle of Port City Stockyards remembers well those days at the end of World War II when cattle were first transported by air. "We didn't have it down to much of a science then," he recalls. "It was mostly hit or miss, and some of those early flights were really risky."

In 1948, Port City Stockyards achieved considerable national attention in the air export field with a large air shipment of livestock to Guatemala City, Guatemala. Two hundred ninety-one head of cattle were transported in consecutive shipments along with horses and other animals.

The project was initiated by Dr. E. J. Kyle, U. S. Ambassador to Guatemala and former Dean of Agriculture at Texas A&M. Kyle wanted to bring a sizeable number of U.S. purebred cattle into the interior of Guatemala without having to rely upon the railroad or an overland, arduous trail drive. He contacted a close friend, Bob Kleberg of the King Ranch, and arranged to have a herd of Santa Gertrudis purebred cattle air-shipped to Guatemala City. The Hudgins Ranch was also contacted for the purchase and shipment of approximately forty registered Brahmans. Purchased mostly by the Guatemalan government, a great fanfare was created when these animals began to arrive in Guatemala City.* [7]

For approximately twenty-five years, beginning in the early 1950's, probably 85% of the Brahmans and other cattle that were exported from the United States were handled by the Houston Freight Forwarding Company. This company's capable and genial president, George O. Puig, remembers that during these years the Venezuelans were involved in a bootstrap program in which they were buying approximately 10,000 head of cattle each year. "The largest percentage were purebred Brahmans or Brahman-blooded cattle."

"The demand for Brahmans became so great," remembered Puig, "that I travelled to Cuba to fill our orders. In fact, we were even exporting Brahmans from Cuba while Castro was in power, but when he found out, he had me arrested as I was opposed to his regime." Puig, born in Cuba, however, was able to pull enough strings to get released and returned home to the United States.

*Other Guatemalan buyers of these first Brahmans included Senor Eduardo Alejas and Senor Asturias (J. D. Hudgins Ranch Records, Hungerford, Texas).

"After Castro became well entrenched, he allowed the fine, purebred Brahman herds to be destroyed," recalled Puig. Later Castro began to import Holstein cattle from Canada. "He even purchased a livestock ship from the Clausen Company," Puig said.[8]

During the years that Puig managed the Houston Freight Forwarding Company, he performed a great service for American cattle interests. His firm not only handled the arrangements for transporting the cattle, most of which went by air, but they were also instrumental in finding buyers for the various breeders.[9]

While the transportation of Brahmans and other cattle by air has had many advantages (particularly regarding time and the ability to reach interior destinations) like all modes of travel, it has produced its complicatons and problems. The actual logistics of arranging transfers by air can become very intricate. Brahman breeder and exporter G. T. Stack remembers that on one occasion in 1977 a Costa Rican customer of his had requested a shipment of eleven Brahmans from four different states. Like the famous Joe McCoy, Stack had said he would deliver.* All eleven head were scheduled to be aïr-lifted from Miami. Stack was faced with the problem of quickly finding a truck and driver to collect the animals and a pilot who could, or would, carry only a partial load. (The eleven Brahmans were about 18,000 pounds short of a full load.) After finally collecting all of the cattle, Stack remembered, "I found a fellow who was hauling cattle from Costa Rica to Venezuela and had to stop over in Miami for a plane service. I was able to catch him at the time he came in, and as he would be going from Miami to Costa Rica without any

*Joe McCoy was an early cattle buyer who convinced Texans to trail their herds to Abiline, Kansas beginning in 1867 where they would meet on an equal footing with Northern buyers ". . . and could be undisturbed by mobs or swindling thieves." McCoy was able to keep his word and deliver, thus originated the phrase "The real McCoy." (See *Story of the Great American West,* The Reader's Digest Assoc., Inc. Pleasantville, N.Y.: 1978, p. 276).

In 1980 G.T. Stack, Brahman breeder, exporter and first president of the Florida International Agribusiness Trade Council, received an award for being the "International Marketing Man of the Year." Here he is shown receiving this plaque from Doyle Conner, Florida's Commissioner of Agriculture. Through the years Stack has travelled to every part of the globe to promote American Brahmans and other livestock.

load, he was willing to haul a 20,000 pound load for a reasonable price."[10]

"More recently," Stack said, "it has been possible to purchase portable disposable crates that run about $450.00 each which can be used for air shipments of cattle only involving a partial load." The cattle are loaded in these crates, just as one might load a large generator or some other equipment, and the loaded crate is placed on a plane by rollers. In Atlanta, trucks carrying the cattle in crates can pull up next to a plane where the crates are then rolled aboard. "It's all part of the evolution of containerization," Stack quipped.[11]

From Props to Jets

While the sturdy and dependable C-47's were employed in the early months of transporting cattle, their maximum payload was under 7,000 pounds. Soon the exporters were switching over to the C-46's which could carry almost double the tonnage of the C-47's. The DC-4 was next utilized which could transport approximately 20,000 pounds. And as George Puig recalled, "When we started using those big DC-6's, we could haul a payload of 30,000 pounds, and we were soon using the Convair 808 which could handle up to 48,000 pounds. It was a prop-jet."[12]

The largest air shipments of cattle have been handled by the giant "stretch jets," which are modified DC-8 jets. These shipments began in the early 1970's. In 1971, J. D. Hudgins, Inc. had a single load of 225 purebred Brahmans air-shipped all the way to the Malagasy Republic - some 12,000 miles distance from their origination in Texas. This total shipment weighed between 90,000 and 100,000 pounds, and it took just 27 hours to complete the transfer.[13]

From the end of World War II until the late 1970's, foreign transfers were extremely important to the American Brahman industry. In 1955, for instance, almost 43% of all purebred Brahmans sold were exported, and in 1959, nearly 50% of all purebred Brahmans sold were exported. It has been an important part of the cattle business, and although cattle exports from the U.S. have been down, hopefully the foreign market will regain its former momentum. It is important to the cattle business as well as to the nation's economy. Indeed, the increased sales of livestock and other agricultural goods are one of the best ways

The first of seventy-four registered Brahman bulls are being loaded aboard a Trans-International Airline DC-8 "stretch jet" bound for Malagasy. The 12,000 mile flight to this island republic took twenty-four hours with four intermediate stops.

for the U.S. to obtain a healthy trade balance, particularly in light of the large amount of oil that is now being imported by the U.S.[14]

Importantly, the sale of cattle to other countries has a snowball effect. As one Brahman breeder put it, "When foreign buyers come to the U.S. they always bring empty suitcases which they take back with purchases they have made here. Foreign cattle buyers usually end up buying other livestock and needed farm equipment. This was clearly demonstrated recently when a party of Panamanian buyers came to Florida to purchase 220 head of purebred Brahmans. While they were here, they also bought 22 registered quarterhorses and left with "filled suitcases."[15]

In the years ahead, cooperation must be aggressively encouraged between the various volunteers, private and govern-

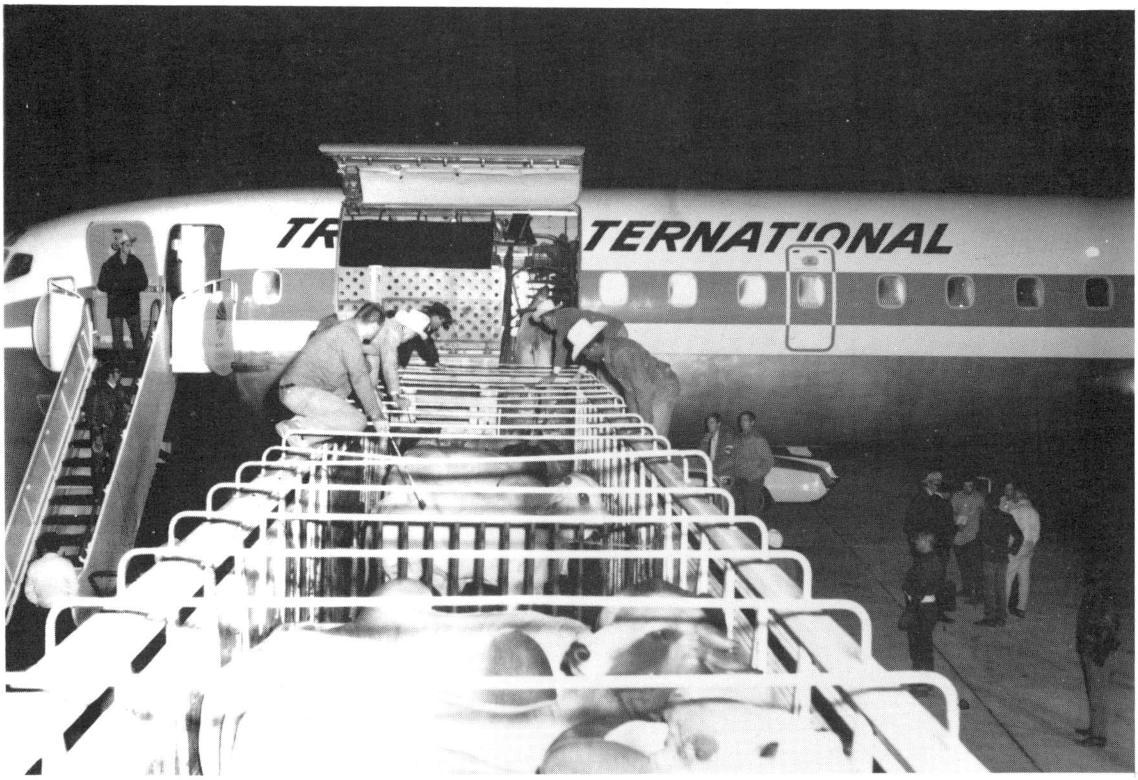

The handling and transportation of Brahmans has been greatly facilitated and modernized since those first flights in 1946. These Brahmans are being loaded on an especially equipped "stretch jet" and are bound for Latin America.

ment agencies, such as the USDA's Foreign Agricultural Service, the U.S. Meat Export Federation, and the various breed associations. The development of air transportation has played an extremely important role in facilitating the transfer of cattle to other countries, but one of the greatest needs today for American cattle exporters is to make other nationals aware of the advantages of buying purebred Amercian cattle. As one Brahman breeder recently explained, "We need to all start working together to promote our product which can help our own cattle industry as well as those who buy the cattle."[16]

10

A Breed for All Seasons

The history of the American cattle industry has been a spectacular one - as dramatic and as romantic as any story unrestrained by truth. Its conflicts and characters have provided the inspiration and literary material for a great American epic.

One of the most amazing aspects of this history has been the speed with which the cattle industry evolved in North America after centuries of near neglect. Nowhere does this seem to be more sharply illustrated than in the history of the American Brahman. In less than a century, this American creation, the product of two widely divergent civilizations, literally changed the ideas about cattle raising in a large part of the world. Because of its economic prepotency, it first won favor among American Gulf Coastal cattlemen for crossbreeding, upgrading existing herds and establishing new breeds. But its popularity was never confined to one region for very long. Where it was introduced, it was generally accepted either for establishing purebred herds or for crossbreeding purposes. Today, Brahman purebred and/or crossbred herds may be found in almost every state of the United States and in nearly seventy different countries outside of the United States. It has indeed become "A Breed for All Seasons."

Early Exports of American Brahmans

Because of its great versatility, its marvelous adaptability and its ability to produce outstanding offspring, it has through the years become an important export item. Despite its superior qualities, the first registered American Brahmans probably were not actually exported until 1933.* During that year, five registered cows and five registered bulls from J. D. Hudgins, Inc. were shipped by rail to the ranch of Mexicans Manuel Guzman Willis and Carlos Guzman Willis at Salinas, Vera Cruz. Some of these cows were later crossed with Guzerat and Nellore type bulls that Guzman had acquired from a 1923 Brazilian Importation. "Having obtained magnificent results, better than the Brazilian types," he continued to purchase Brahmans from the Hudgins Ranch at Hungerford, Texas.[1]

In 1936 Guzman was commissioned by the President of the Republic of Mexico to buy more American Brahmans from the United States. The Mexican President was particularly impressed with the American Brahmans as an American Brahman that Guzman had purchased for him earlier had won a first prize at the Houston Fat Stock Show in 1935. Interestingly, 1935 was the same year that Guzman and his brother enrolled in the ABBA, making them the first Mexican citizens to do so. Also, that year Walter Hudgins of J. D. Hudgins, Inc., along with some other Americans, inspected the Guzman Willis ranch giving a "AA" classification to the great bull *El Conejo*, ABBA 1020. *El Conejo* was a fifteen-year-old "pure" Guzerat from India, and was considered to be the nucleus of the Guzman Willis cattle herd.[2]

*While ABBA registered cattle were not exported until this time, the Lykes Brothers had begun shipping Brahman cattle to their ranches in Cuba in the 1920's from the port at New Orleans. Transferred by Lykes' ships, some of these cattle were later registered in Cuba. Mr. R. E. Thompson of Wilson, Louisiana remembers crating high-grade Brahmans that went to the Lykes and then to Cuba in the early 1920's. (Harry Gayden, former ABBA Executive Secretary, interview, Oct., 1980, Dallas, Texas.)

Bos indicus **cattle were probably first imported into Mexico about 1880. By 1935, registered American Brahmans were being brought in from Texas. This Brahman bull, owned by J.T. White of Hearne, Texas, was later used as a herd sire on a ranch in San Luis Potosi, Mexico.**

In 1933 the first American Brahmans were exported to Australia. In the beginning, the Australians were faced with the same problem that American coastal cattlemen had faced earlier -that was to produce a beefy type animal that would have a carcass pleasing to the average housewife and at the same time be able to thrive within a harsh environment.[3] Since those early beginnings, the Australians have incorporated many Brahmans into their beef programs, and today purebred Brahmans are being grown in most parts of Australia, including the provinces of Queensland, New South Wales, Victoria, Western Australia and the Northern Territory. Of the approximately 20 million cattle in Australia, it is estimated that at least one-half carry Brahman blood.[4]

In 1939, two Brahman bulls from the Dr. W. S. Jacob's herd were bought by E. P. Halliburton, a Houston oilman, and exported to the President and the Secretary of Agriculture of Honduras.[5]

During the 'thirties American Brahmans were also exported from the United States to Colombia, Costa Rica, the Fiji Islands, Puerto Rico, Tahiti, Canada and Venezuela.[6]

Growing Popularity in States

The last years of the 'thirties were marked also by the growing popularity of the Brahman in other parts of the States as more and more breeders began to use them in North Texas, Florida, Louisiana, Mississippi, Arkansas and other regions of the Gulf Coast area.

Near Kaplan, Louisiana, one of the Coastal Region's early Brahman breeders, Aubrey J. Marceaux, was establishing a reputable herd with cattle purchased from the Sartwelle Brothers of Palacios, Texas. In 1934, Marceaux bought a three-year-old bull from the Sartwelles out of the incomparable *Imperator*, ABBA 120. Marceaux's new bull was named *King of Kaplan*, destined to be one of the greatest bayou bulls of all times. Marceaux also purchased twelve other cattle from the Sartwelles out of the *Aristocrata*, ABBA 134, line. In 1941, Marceaux also added *Aristo Rath*, ABBA 7058, by *Aristocrata*, out of an *Imperator* daughter to his herd.[7]

Always anxious to improve and to promote the American Brahman breed, Marceaux took a direct role in the early Brahman studies that were conducted at Louisiana State University by furnishing the station with outside blooded Brahmans descended from *King of Kaplan*.

Rasputin (Imperator of LSU 2-206), son of *King of Kaplan* was one of the best sires at L.S.U.* Marceaux later borrowed *Rasputin* and bred him back to the *Imperator* line, getting the *Mac Marso 670* bull. One of the best herd bulls of the Marceaux Ranch was the *AJM Imperator Marso 990.* Although this bull has been dead for several years now, the Marceauxs still use his semen for breeding. One of their important herd bulls today is *AJM Imperator Marso 1/7*, son of the 990 bull.[8]

Alphonse Forbes of Greenwell Springs, Louisiana was another Louisiana breeder who was laboring to establish a reputable Brahman herd during the late 'thirties and 'forties. Forbes was buying his foundation cattle from the Pierce Estate, the Sartwelle Brothers, Dr. W. S. Jacobs and C. C. Appling.

Herman and Margie Rogillio (daughter of Alphonse Forbes), owners of the Forbes Ranch today, are currently involved in an F_1 Brahman program involving both Hereford and Angus cattle. Margie Rogillio reports that they now have fifty F_1 Brahman mother cows that have been certified through the ABBA with twenty-seven more F_1 females on the ground. The Forbes' *Bevo* bull, *Bevo Repucho* 726, purchased from Herman Beville of Bushnell, Florida, *JJ Didor Imperator 269*, and *McKellar Arauto* are their prize herd sires.[9]

Besides the fact that many fine Brahman herds were being established in Louisiana during the 'thirties and 'forties, the great Bayou state also had the distinction of being the site of the first

* *Rasputin* was by *Imperator Manso*, PH 699, ABBA 20883, a son of *King of Kaplan* out of the Sartwelle cow *Commander's Rose*, PH 81, ABBA 13051, a daughter of *Imperator II*, PH 327, ABBA 3001.

National Brahman Show.* Held at L.S.U.'s Agricultural Center in 1946, it was organized and sponsored by the Louisiana Brahman Association. This event was followed the same year by the first Brahman Field Day, held at the ranch of J. M. McLemore of Alexandria, Louisiana. Three years later the first non-Texan, Herman Taylor of Louisiana, was elected the president of the ABBA. One of the active affiliate associations of the ABBA has been the Louisiana Brahman Association which recently elected Elmer Lolley of Winnsboro, Louisiana as their new president.[10]

During the late 'thirties and early 'forties many Florida cattle breeders also began to associate themselves with the best breeders in Texas, ". . . often times finding the cow in one herd and the bull in another, in order that they might have the very best of breeding stock in many herds in the State of Florida."[11]

In 1944, a group of Florida's leading Brahman breeders met at Arcadia (once known as 'Tater Hill Bluff'*) in Florida ". . . to make a united effort to get as many as they could of the best of the Brahmans. A year later, they put on their first show at Ocala, Florida . . . the largest show that had ever been held in the world of registered Brahman cattle." With the introduction of Brahmans, Florida became in the 1940's the second largest beef producing state in the South, exceeded only by Texas.[12]

The Florida Brahman Breeders Association also sponsored three early public sales in 1945, 1946 and 1947 at Ocala,

**Manimoso Manso*, a *Manso* lined bull, was the Grand Champion at this show. After the show, Edgar Brown of Orange, Texas, offered the then unheard price of $20,000 to the Hudgins Ranch for the bull; however, the Hudgins refused to sell. (Edgar Hudgins, Hungerford, Texas: March 20, 1981.)

*At one time Arcadia was one of the wildest cow towns in the United States. During the late 1800's several cattle trails, leading to embarkation points on Florida's Gulf Coast, passed near Arcadia. (See Joe Akerman, "Life on the Florida Frontier," *Florida Cowman*. Florida Cattleman's Assoc., Kissimmee: 1976).

Florida. Commenting on the quality of Brahmans at the 1947 Ocala Show, Henry O. Partin, one of the cattle deans of Florida's modern cattle industry, said, "This sale offering will be gone over by the Committee like a mother-in-law's tongue reviewing a son-in-law's business." Dr. Carlos Smith, a Brazilian breeder, who visited the first Ocala show in 1945, ". . . spoke of the beef qualifications and the size and wonderful development of these Florida cattle," expressing pride that many of these cattle were blooded to cattle from his own home at Uberaba, Brazil. Buyers representing the government of Dutch Guiana also purchased some $14,000 worth of Florida Brahmans from members of the Florida Brahman Breeders Association at this show. And as early as the 1940's, approximately thirty carloads of Florida crossbred-Brahman cattle were purchased by breeders from Arizona, including Wirt G. Bowman of Nogales, Arizona.[13]

Some of the leading Florida purebred Brahman breeders at this time included the Partins of Kissimmee; the Stuart Brothers of Bartow; the Bevilles of Bushnell; the Norris Cattle Company of Ocala; Milton R. Thomas and Sons of Sarasota; I. A. Krusen of Zephyrhills; Frank Bass of Wauchula; Crum Cattle Company of Plant City; the T. J. Durrance and Son Ranch of Brighton; A Duda and Sons of Oviedo; John Turner and H. M. Lanier of Arcadia; the Monticello Drug Company of Jacksonville; the Caloosa Ranch of Okeechobee; the United States Sugar Corporation of Clewiston; E. Vose Babcock, Jr. of Fort Myers; Mrs. Pat Johnson of Kissimmee; Cushman S. Radebaugh and Sons of Orlando; The Whaley Company of St. Cloud; T. M. Britt of Winter Garden; R. C. Bentley, T. C. McKown and Lamar Beauchamp of the Polk Brahman Farm of Winter Haven; Wilbanks Brothers of Tampa; Dr. T. P. Chaires, Jr. of Bradenton; Clyde J. Keys of St. Petersburg; L. S. Harris (President of ABBA, 1958-1960) of Kissimmee; A. Y. Carlton, Sarasota; Lester Brothers of St. Petersburg; Herschel A. Auxier of Miami Beach; and William R. Boyd of Lake Wales.[14]

Fairly typical of Florida's early Brahman breeders is Herman Beville of Bushnell, Florida. Beville's family has raised beef cattle in Florida since the 1850's when his great-grandfather, Granville Beville, came down the old Beville Trail from Georgia with a large herd of wild woods cows. Beville recently recalled the following:

> *All anyone had then were 'ole woods cows, but there was plenty of them. About 1915, the Lykes started bringing a few Brahmans into this part of the state. They were about as good a looking cattle as had been in this country.*
>
> *In the early 1920's my great-uncle started with a few grade-Brahmans and was real happy with them. Of course, at the time there was no breed association or anything like purebred Brahmans. It was later on in the 1930's that I went to Texas and bought some purebreds from Cecil Boyt of Devers, Texas. Later on we also picked up some fine bulls from the Sartwelles, the Hudgins and Raleigh Sanborn of Bay City, Texas. And we also later bought a bull from the Partins of Kissimmee. It was out of the Emperor bull. I also went to Texas a number of times and brought back several carloads of Brahmans. Two of my best early bulls were Bevil B. Manso 88/4 and Greta II Imperator, 0/855.*
>
> *Since we started in the Brahman business, we have come a long ways and established one of the finest Brahman bloodlines around.*[15]

Of the numerous Texas breeders who were expanding, establishing or upgrading Brahman herds at this time, one was Vernon Frost, whose Pecan Acres Ranch is located at Simonton, Texas near Houston. Frost, a grandson of J. M. Frost, who had brought some of the first *Bos indicus* cattle to Texas in the 1880's, had owned only purebred Brahman bulls until about 1941. That year he purchased some females from Alma Boyd, and in 1946, he bought some of the *Manso* bulls and 40 purebred cows from the Hudgins Ranch. In time he had developed an outstanding Brahman herd. "At that time," as Frost remembered, "I was grazing my cattle where the middle of Houston is today." Vernon Frost's older brother, J. M. Frost, Jr., has also been one of Texas' most prominent Brahman breeders. From 1945 to 1947 he served as the ABBA's fifth president.

During the 1940's, Wayne Fisher, Brahman breeder and meat packer from Blythe, California, vigoriously strove to popularize

Purebred Brahman cattle graze peacefully on mile high pasture in the great Southwest. This picture was taken near Reno, Nevada.

the important attributes of the American Brahman among Pacific Coast and Far Western cattlemen.* During one week of 1946, ". . . registered Brahman cattle rolled from the Fisher Ranch . . . to four Western states and the Hawaiian Islands." The largest shipment of these Brahmans went to breeder, packer Paul Cornelius who put them on his ranch near Reno, Nevada. Grazing on a range framed by snow capped mountains, it was a different scene for people who were used to seeing Brahmans in a tropical or semi-tropical setting.[16]

Mrs. Lee Henderson, who was using Brahman bulls during these years in the arid region of New Mexico, where temperature

*Fisher was once cited by the ABBA for the outstanding work he had done in the Far West in publicizing the American Brahman. (See ABBA minutes, 1924-1980, ABBA office, Houston, Texas.)

extremes were common, reported that during colder weather her bulls put on longer coats of hair and ". . . wintered surprisingly well."[17]

While Brahmans seemed to be catching on when first introduced to the Far West, by the early 'fifties, this early demand had practically ceased after a number of poor-grade Brahmans were brought into the area by speculators. However, as a few dedicated breeders began to upgrade their Brahman herds, their popularity began to return. Prices brought by high quality purebred Brahmans and Brahman-crossbreds have remained consistently high during the last decade in California.[18]

About 1952, Paul Cornelius of Coleville, California, bought out the Fisher herd and some of these were eventually sold to breeder Avon Davis of Chowchilla, California. Davis, who has been promoting Brahmans for about forty years, developed a polled Brahman herd from the Fisher herd with the addition of a *Repucho* blooded bull he purchased from Mrs. Wilbur Webb of *El Campo, Texas.*

Davis says that he anticipates a good future for Brahmans and Brahman-crosses on the West Coast. In a recent interview he said:

> *A good number of people in these parts like Brahmans for producing Brangus cattle, and of course, many are aware of the advantages of the* F_1 Brahman cow.
>
> *One problem with these cattlemen in the past has been that they didn't want to have anything to do with cattle that didn't have white faces - even though pinkeye and other diseases were common and could be cured with some Brahman blood. But a good number of the younger breeders are now starting to mix Brahman cattle with their herds.*[19]

By the late 1970's there was enough interest in the American Brahman in the Far West for the creation of a Brahman association, and in 1979 the Western Brahman Breeders Association was established which became affiliated with the ABBA. Today there are 45 active Brahman breeders in this association representing the states of Arizona, California, Nevada, New Mexico and Oregon.[20]

Certainly the principal indicator for judging the acceptance of a cattle breed is how well it sells when it is put in competition with superior cattle from other breeds. In 1979, Brahmans did outstandingly well at two annual sales held in California. One was the Famosa Sale held near Bakersfield, and the other was a sale held at Red Bluff located on the northern reaches of the Sacramento River. In both of these sales, American Brahmans topped all other sales. At the Famosa sales, Brahman bulls averaged $2,500, most of them coming from Texas. The famous San Francisco Cow Palace Sale was also topped in 1979 by a Texas Brahman bull, *Mr. 3X Mucho 62,* which was also the Grand Champion bull in the sale among all breeds. At an all breeds California sale held in January, 1981, registered Brahmans averaged approximately $2,200 per head while Brangus averaged $1,500 per head and Shorthorn averaged $1,300 per head.[21]

In the arid Southwest at Tucson, Arizona, where the annual Southwest Bull Classic was held in March of 1980, thirty-seven Brahman bulls topped all sales, averaging $1,850. This was an average of $504 more per animal than the next highest breed average. Three Brahman bulls consigned by Diamond T. Ranch, Bay City, Texas; J. D. Hudgins, Inc., Hungerford, Texas and Frank Connell of Thornton, Texas, all received a bid of $2,700 to tie for high selling honors at this sale. In 1979, the Geyser Ranch of Ely, Nevada purchased 9 bulls at the Southwest Bull Classic sale, paying a total of $24,000.[22]

In the show that preceded the 1980 sale, the grand champion bull was *ADS Duson's Best 846/3,* owned by Jerry King of Hattiesburg, Mississippi; while the grand champion female was *LMT Miss Diamond T 241,* owned by the Diamond T. Ranch of Bay *City, Texas. The reserve champion bull was JDH Jef Rhax Manso,* owned by J. D. Hudgins, Inc.[23]

While the acceptance of Brahman and Brahman-crosses appears to be growing in the Far West, Brahmans have also been tested in other parts of the country and have done well at various

This Brahman bull, *A D Repucho 144*, **ABBA 25-3289, owned by Avon Davis of Chowchilla, California, was used for breeding in Washington State for three years. He is seen here on the Davis Ranch with Vinnita Davis.**

times in a number of places not normally associated with Brahman raising. Near Boyce, Virginia, for instance, cattleman and farmer Brooks Mims reported in 1951:

> *For our markets, the best animal to hang as a carcass on the hook is the cross [Brahman] with one of the English breeds. . . .*
>
> *[Brahmans] give more beef per acre. At weaning the calves go 75 to 150 pounds more than those of other beef. . . the steers will go to the butcher right off the grass, and the calves can be weaned on grass alone and not be stunted.*[24]

During the same year, the R. D. Littles, the Frank G. Smiths and the William Millers, all Brahman breeders in Ohio, entered

Brahmans for the first time in the Ohio State Fair at Columbus. Although cattle from the Norris Cattle Company of Ocala, Florida won most of the ribbons in the Brahman classes, the Ohio breeders were no less enthusiastic about their Brahmans. All were satisfied about the way these cattle withstood the heat -as well as the cold, and because of local interest, the Central Brahman Breeders Association was organized with William H. Miller as its first president.*[25]

One of the most active young affiliate associations has been the Carolinas Brahman Association. Lewis Patterson of Concord, North Carolina, who has been raising Brahmans for over 34 years, has been one of this association's driving forces. Begining with *Imperator* and *Manso* line cattle from Florida, Patterson has established an outstanding herd and has sold registered Brahmans to buyers in states all over the Northeast, including breeders in New York State and Massachusetts. The Carolinas Brahman Association holds two annual field days, one at Lugoff, South Carolina and the other at Monroe, North Carolina. In July, 1981, the Carolina Association sponsored its first sale at Lugoff, South Carolina. The current president is John Coble of Concord, North Carolina.[26]

The growth in popularity of the American Brahman in the United States can be seen further by the fact that the following states have affiliate associations of the American Brahman Breeders Association: Alabama, Arkansas, Florida, Georgia, California, Mississippi, Louisiana, Oklahoma, Tennessee, and North and South Carolina. Besides these state affiliates, there are also nine other regional associations. Most of these are in Texas and Louisiana.

The Expanded Use of Brahmans in Latin America

The first domesticated cattle to reach the Western Hemisphere were brought to the West Indian island of Hispaniola by Colum-

*This assocication is no longer active.

bus in 1493. These hardy cattle, conditioned by years of environmental extremes, were descendants of the old landraces which had been domesticated in Spain for centuries. In the New World these Iberian scions would come to be known as Criollo cattle. (See Chapter I, "What Is the American Brahman?")

Cattle from this tropical island were soon brought to other islands of the West Indies and eventually to the mainland by adventurous conquistadors and devout ecclesiastics. Herds grew in such abundance in some areas, such as Cuba and Mexico, that they were soon running wild as did the native buffalo or later as did the Texas Longhorns and the scrub cattle of the southern coastal regions.[27]

After the beginning of the nineteenth century, importations of modern European breeds began in several parts of Latin America as the Latin Americans attempted to upgrade their herds with new blood. Important to our study, however, is the fact that by the mid-nineteenth century, *Bos indicus* cattle were being brought to Jamaica, and perhaps other West Indian islands, directly from India. Their adaptable and heterotic qualities were quickly recognized, and their use for crossbreeding quickly spread to continental Latin America. By 1870, sizeable numbers of Indian cattle were being imported by Brazil. Descendants of these cattle were soon being dispersed over other parts of Latin America and in most cases crossed with native cattle.[28]

As tick-borne diseases began to take their greatest toll from the native stock of Latin America, more and more ganaderos began to incorporate *Bos indicus* cattle into their herds. A few years before the onset of World War II, American Brahmans began to show up in Cuba, Colombia, Puerto Rico and Venezuela. They quickly became popular as they demonstrated all the desirable qualities of other *Bos indicus* breeds as well as having the beef-like qualities of the European breeds. "After World War II, the American Brahman practically replaced the Indian Zebu in breeding out the Criollo herds throughout Latin America, except in Brazil."[29]

Pictured is North Carolina Brahman breeder Lewis Patterson with one of his fine herd sires at his ranch near Concord, North Carolina.

The number of registered American Brahmans coming into Cuba made a dramatic increase during the 1940's. And in 1944, ABBA membership had grown to such a degree in Cuba that the Cubans requested a special by-law from the ABBA Board of Directors regarding the inspection of Brahmans in Cuba submitted for registration. This by-law required that special inspection committees be selected by the ABBA Directors as inspection teams. Almost all of these members were from the United States. By 1953, ten percent of the then 199,058 Brahmans registered by the ABBA were on Cuban ranches. And during the

This Cuban vaquero is seen holding an American Brahman shown at the *St. Spiritus Cattle Show* **in 1950. This cow,** *Jisara*, **ABBA 61219, won the ABBA trophy that year.**

1950's, the first International Brahman Congress ever held was hosted by Cuba.[30]

During the 1940's and 1950's, before Castro became dictator, some outstanding Brahman cattle were shown in Cuban shows. At the once popular National Cattle Show held in Havana each year, a registered American Brahman named *Sacuquita* was named Grand Champion Female and Best Grand Champion Female of all breeds in 1947. And at the once famous *Sancti Spiritus* Show in 1946, a beautiful bull named *Piedra Manso* won first prize and over-all champion.[31]

Since the United States broke diplomatic relations with Cuba in 1960, no Brahmans have been registered by Cubans. Existing breeding programs were evidently scrapped. One former manager of the King Ranch's Cuban operation recently said that the best American Brahman and Santa Gertrudis cattle were exported to the Soviet Union after Castro seized power in 1959.[32]

One of the largest and most successful American Brahman operations in the West Indies is presently located on the Gulf and Western's Central Romana Ranch in the Dominican Republic. "In this tropical setting of luxury and exciting voodoo backdrops . . .," Gulf and Western maintains a herd of approximately 30,000 head of beef cattle. Approximately 1,000 are purebred Brahmans based mostly on Sugarland's and Hudgins' breeding stock, many having been sired by *Sugarland's Rex Manso 48th* or *Sugarland's Dipcrata 7th*, both former National Grand Champions.[33]

Besides their purebred American Brahmans, in 1974 Central Romana was also running 100 registered Santa Gertrudis, 750 registered Charbray and 5,000 purebred Romano Reds. The Romanos, also a *Bos indicus* breed, were actually created at Central Romana fifty-eight years ago.

Central Romana has relied heavily upon a soundly run A. I. (Artifical Insemination) program to perpetuate its better breeding lines. And it has one of the most up-to-date operations in all of Latin America.

Recently one of the truly bright spots for exporters of American Brahmans in the South Caribbean has been the beautiful Windward Island of Martinique. Having an area of only 425 square miles, the first registered American Brahmans were imported by this French island in 1951. But beginning in 1976, small quanities of Brahmans were brought in from the States on a fairly regular basis, and in 1980, fifty-nine American Brahmans were imported.

An ABBA team that recently visited the island reported that livestock development there was particularly good with the evidence of Brahman blood in many cattle.[34]

The Martinique-Brahman connection demonstrates how government concern and available credit can facilitate international trade. The availability of credit through the Agricultural

Pictured here are part of the nearly 1,000 head of ABBA registered Brahmans on Gulf and Western's Central Romana Ranch in the Dominican Republic.

Bank of Martinique and governmental subsidies by the French government have been a great help to Martinique ranchers who are striving to upgrade their herds.[35]

In South America the American Brahman has helped tremendously in upgrading many existing herds. Nowhere has this been truer than on the vast, open plains of Venezuela and Colombia known as the Orinoco Basin or the Llanos. This great treeless region covers about 300,000 square miles - an area larger than Texas, where hot winds often scorch the grasses in the dry season and heavy rains flood the prairies during the wet season.

Cattle raising was the sole occupation of the pioneers who first trailed Spanish cattle into this area. By natural selection in an oft times adverse environment, there developed the tough, little Llanero or the Criollo of the Llanos. It is as tough a bovine beast as ever felt the lash of a vaquero's whip. But like the Florida scrub, after successive generations it had become small and lack-

Three to four year old American Brahman and Brahman-crossed steers are seen here near Valencia, Venezuela on a ranch managed by the Lancashire General Investment Co., Ltd.

ed few of those qualities desired in beef animals. Its most prized quality remained its ability to endure as it often ranged uncontrolled in nearly a wild state just as the Longhorn had once roamed over the Texas range.[36]

As late as the 1950's, practically all of the cattle of the Llanos were the Llaneros-pure Criollos. Importations of Zebu stock began to increase, coming first from Brazil, but eventually American Brahmans became the most popular breed. And from 1954 to 1963, approximately 17,000 head of purebred American Brahmans were imported by Venezuela. Many were brought to the Orinoco Basin. The results were remarkable as Brahmans were crossed with Llaneros and the best qualities of each were preserved. The heterosis from the first crosses showed a great advantage in size over the native cattle. Purebred Brahman herds were soon established to provide bulls for upgrading. Today most of the herds in the more accessible areas of the Llanos are upgraded American Brahmans, but in the more remote districts, some Criollo cattle still range along with American Brahman-Criollo crossbreds which are still in the process of being upgraded.

As the quality of beef breeds has greatly improved in Venezuela, the cattle operations of the small ranchers and farmers has also shown real improvement. It has generally been in the huge, more remote holdings that cattle are still handled today as they were centuries ago. Ranches are often so large, some over one million acres, that construction of fences is extemely expensive. Cattle are normally identified by brands during roundups just as they were once on North America's frontier. But even in some of these areas roads are being constructed and better management practices are in evidence. While the approximate cattle population for Venezuela is 9,000,000, it is believed that the real potential for producing cattle is yet to be realized. As this potential is more fully realized, the American Brahman should play an even more important role.[37]

Part of the evolution of the Venezuelan beef industry and the demand for American Brahmans has been due to the expansion of credit programs by the government of Venezuela during the last two decades. Import licenses have been very difficult to obtain, however, during the last few years, and imports of American Brahmans since 1979 have practically ceased. One of the main reasons for this appears to have been credit that has been provided for the Venezuelan breeders by the Brazilians. But since the American Brahman has helped Venezuela's cattle production so much during the 1950's, the 1960's and most of the 1970's, surely this trade will be resumed soon.*[38]

*Steve Abecasis, former ranch manager of the British owned Lancashire General Investment Co. Ranch of Venezuela (which runs about 100,000 head of cattle), and Pat Foster, Lancashires' present manager, recently pointed out that the move among breeders in Venezuela was away from the "more featured type *Bos indicus* cattle and more toward the beef type cattle of the American Brahman breed." (Joe Dial, ABBA Director, Placedo, TX: Sept. 6, 1981.)

Colombia, Venezuela's neighbor to the west, is a land of vast pastures and thick forests. The eastern three-fifths of this country is a broad plain, almost at sea level, practically devoid of human population. Its northern region is part of the Llanos that stretches over into Venezuela.

Colombian ranchers apparently did not start utilizing *Bos indicus* cattle until 1913 when some were imported from the West Indies.* They were brought in because of their tolerance to tickborne diseases, but as so few were imported, they actually made little impact on the existing Criollo herds or the cattle of the country.[39]

Colombia got into the business of importing American Brahmans in 1938 and later undertook an ambitious program of upgrading their far-flung herds with the use of American Brahmans and such Brahman derivative breeds as Brangus, Charbray, Santa Gertrudis and Beefmasters. In 1953, more than 4,000 head of breeding cattle were exported from the United States to Colombia - most of which were registered Brahmans.[40]

Since the 1950's, Colombia has continued as a steady customer for exporters of American Brahmans. While imports peaked during the 1950's, the demand for high grade Brahmans began to increase again in 1977. And since then, approximately 1,200 registered American Brahmans have been imported by the Colombians as they have maintained their preference for this breed. According to one American breeder, "Probably more top quality, high priced American Brahmans have been exported to Colombia than to any other country." These imports have made a dramatic improvement on native herds.[41]

For its size, today Colombia has one of the largest cattle herds in the world, approximately 24,000,000. Most of these are native cattle carrying a strong Brahman influence or purebred

*John Rouse in *The Criollo, Spanish Cattle in America*, University of Okla. Press, Norman: 1977, states that some Zebu cattle were actually received as early as 1908 from Puerto Rico and Texas. p. 136.

and grade Brahmans with varying amounts of European and Brahman-derivative breeds. Even in the more isolated areas of the mountain ranges and the eastern plains, the Brahman influence is evident. Some Colombian breeders of American Brahmans have built up their herds through the years to such a degree that they have been exporting numbers of registered and grade Brahmans to other Latin American countries.[42]

Despite the size of Colombia's national cattle herd, the potential for increased beef production in this Latin American republic still appears almost unlimited. One important reason for this is the underutilization of thousands of acres of good grasslands.

Ranching on some remote parts of the great Eastern Colombian plains is similar today to the style of ranching that prevailed in America's West during the nineteenth century. Roundups for branding and the castration of males is often held once a year, but the sole duty of the vaqueros after the roundup is usually to keep track of the herds on the open ranges and to prevent cattle rustling.[43]

While nostalgic examples of such management still exist today, they are gradually changing, particularly where areas have been made more accessable by new or improved roads. In recent years there has also been an increase in the practice of bringing cattle to improved pastures before marketing.[44]

While better management practices are being implemented in Colombia, a demand for American Brahmans and other Brahman derivative breeds will undoubtedly continue. As is true in all countries, however, political, economic and social conditions will play an important part in the future of the livestock industry in Colombia.[45]

In terms of importation numbers and the total impact on the once existing cattle industry, Argentina is another Latin American country that should quickly come to one's mind when considering the American Brahman.

In this romantic land of the unvanquished gaucho and the rolling, sea-like pampas, the advent of cattle raising began about 1570 in the northern lowlands - a land of savannah broken only by occasional scrub forests. It was over a century later before wild cattle were hunted for their hides on the Pampas by the free-wheeling, wide-ranging gauchos. Cattle that ranged over these vast plains also spread to the Uruguayan grasslands - some of the best pasture in Latin America.*[46]

The first non-Spanish cattle came into Argentina in the 1840's from Britain, and by the 1870's Shorthorn bulls were being used extensively with Criollo herds. Their use was followed by Hereford and Angus breeds. By the middle of the twentieth century, British breeds dominated the Pampas. But in semi-tropical northern Argentina, Zebu cattle from Brazil were being crossed with the Criollo herds.[47]

While Zebu cattle from Brazil continued to enter Argentina, in 1941 the first registered American Brahmans were imported from the United States. In some areas they caught on fast. And from 1969 to 1980 nearly 3,000 registered American Brahmans entered Argentina. Interest has grown so much during the past decades that a national Brahman show is now held annually at Virasaro, state of Corrientes, in Northern Argentina. Brahmans are also shown at the Palermo Show in Buenos Aires, Argentina.[48]

In the last few decades, the cattle industry of Argentina has made phenomenal progress in the development of better quality stock and the marketing of younger animals. In some areas an important part of this progress has been due to the use of the American Brahman and the Santa Gertrudis breeds.[49] Most of the ranchos of Argentina are extremely large, and the greatest potential for increased beef production appears to be in the use of more crossbreeding with Brahman cattle and the greater utilization of grassland. One of Argentina's major problems today is inflation which has had an adverse effect on meat packers.[50]

The size of the national cattle herd is approximately 45,000,000.[51]

* The importance of hides in Argentina during the late eighteenth century can be seen in a customs report released from Buenos Aires in 1783 which accounted for 1,400,000 hides. (See John E. Rouse, *The Criollo, Spanish Cattle in the Americas*, University of Oklahoma Press, Norman: 1977.)

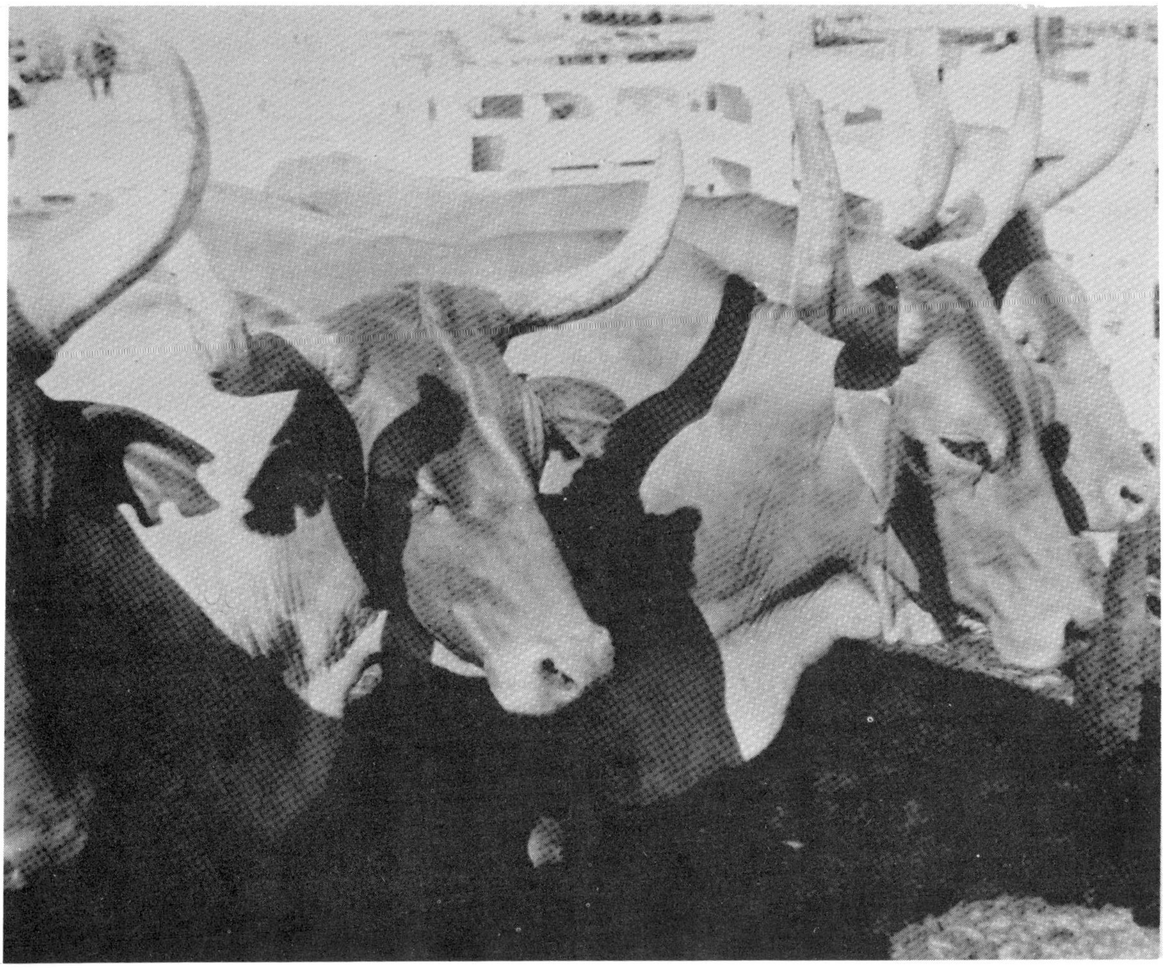

These are Brahman-Criollo crossbred steers from northern Argentina at a Buenos Aires stockyard in 1964 (courtesy John E. Rouse, *The Criollo, Spanish Cattle in the Americas*, **Univ. of Oklahoma Press: Norman, 1977, p. 96)**

While South America's largest country, Brazil, has not utilized American Brahmans as such, the Brazilians have been in the *Bos indicus* business for over a century. And, of course, a number of important shipments of *Bos indicus* cattle have been exported to the United States and Mexico. (See Chapter IV, "Those Brazilian Brahmans.")

As mentioned earlier, American Brahmans are not accepted as purebred animals by Brazilian interests which is doubly ironic as the ABBA does not register *Bos indicus* cattle unless they are judged to be at least 31/32 pure, a stronger requirement than most other registries demand. And since registered American Brahmans have reproduced consistent phenotypic types for over 50 years, the question of purity was settled long ago.[52]

Originally stocked with Portuguese cattle that also developed into a Criollo type, Brazil started importing *Bos indicus* cattle from India in the 1870's. Many importations followed and today most Brazilian cattle herds are Zebu or Zebu-related cattle types. The only real exception are the dairy herds near the population centers which are generally European breeds.

Some Brazilian breeders have maintained a purity of breeds from original Indian stock - separating Nellore, Gir and Guzerat breeds. A fourth distinctive breed developed in Brazil is the Indu-Brazil which was created by combining these breeds. The latter is a more rangy animal.[53]

It has been suggested that if Brazil fully exploited its cattle production, it could move away from its one crop economy - coffee.[54] While Brazil's national herd has been estimated as high as 91,000,000 head, many cattle are trailed over vast distances during the year to new pastures. It is also believed that large areas of new pastureland could be created without effecting present cropping needs.[55] It goes without saying that the importation and use of American Brahmans would be a real help to Brazilian breeders for improving and expanding present herds.[56]

Some of the largest cattle operations today in South America are on the historical Gran Chaco. This extensive low lying plain covers more than 200,000 square miles. More than half of the Chaco belongs to Paraguay, but parts of it extend into northern Argentina and southern Bolivia.* The Chaco has many great

*It has been estimated that 25 million head of new cattle could be fed in the north-central region of the Beni region of Bolivia. (See Loren McIntyre. "Flamboyant Is the Word for Bolivia." *National Geographic*, February, 1966.)

swamps as well as thorny brush jungles and grassy plains. And it is an area where Brahman and Brahman-blooded cattle have practically replaced Criollo herds and European-crossed animals. With the introduction of Brahman cattle, resistance to insects and heat stress has been substantially increased, and better weights have been consistently obtained on steers destined for slaughter.[57]

Imports of registered American Brahmans and Santa Gertrudis cattle have been fairly consistent into this area, particularly into Paraguay. During 1980 over 120 head of purebred American Brahmans were sent by air freight to Paraguay.[58]

Much of the Gran Chaco where cattle are raised is also similar to the early American frontier. Here horses and oxen are sometimes the chief mode of transportation, and commercial cattle operations may run from 25,000 to 100,000 acres. Some estancias are so isolated that cattle sometimes have to be trailed two or three weeks before reaching market. Interestingly, one of the chief predators of calves on the large estancias are jaguars, or *tigres* as they are locally called. Hunters are sometimes employed to kill the *tigres*. "For his food and the use of three horses, he spends his time hunting in the Chaco and is seldom seen around the headquarters. His return is thirty to forty pelts a year at $50 each."[59]

There are few areas of South America that have not been influenced by the American Brahman, its derivative breeds or other Zebu cattle. Whereas not all South American nations have the large cattle herds that populate Venezuela, Colombia, Argentina, Brazil or the countries of the Gran Chaco, other countries, such as Ecuador, Surinam and parts of Peru and Chile also have the potential for expanding beef production with the use of Brahman cattle. Ecuador, for instance, whose Branman breeders registered 573 cattle with the ABBA in 1972, has been a fairly steady importer of registered American Brahmans, and its government is presently pursuing a public program to import up to 63,000 head of cattle. One of Ecuador's outstanding Brahman breeders is ABBA Director Leon Febres-Cordero of Guayaquil. The creation of a liberal loan program by the

American Brahman-Criollo crossbred heifers at an Ecuadorian estancia in the Costa, Ecuador (courtesy John E. Rouse, *The Criollo Spanish Cattle in the Americas*, **Univ. of Oklahoma Press: Norman, 1977, p. 106.)**

Ecuadorian Ministry of Agriculture should help to make all this possible. "Imported animals are generally distributed to qualified livestock producers under 9-year loans including 2 years' grace."

Recent ABBA teams travelling to Ecuador which have included Sloan Williams, Dr. Richard Forgason, Joe Dial and Wendell Schronk indicated that there is a real interest there in American Brahmans with a good market potential for the future.* [60] Over 150 people recently attended a Brahman seminar held at Guayaquil, Ecuador that was sponsored by the ABBA and the Asociacion de Ganaderos del Litoral of Ecuador.

While Surinam (also known as Dutch Guiana) began importing American Brahmans in 1946, financing problems seem to provide the greatest obstacle today as far as the importation of American cattle. An ABBA team that visited Surinam in 1976 made up of Directors John Watson and James Coker reported that there was an interest there in bringing in more registered American Brahmans. Some of the commercial herds in Surinam are now running from 2,000 to 8,000 head.[61]

While *Bos indicus* cattle were probably not brought into Central America until the early twentieth century, since their introduction, they have had an extraordinary impact on this region's cattle industry. This has been particularly true since the importations of American Brahmans began in the late 1930's.

Costa Rica, which had first been stocked with Spanish cattle from Panama during the sixteenth century, was the first Central American country to import sizeable quantities of registered American Brahmans. As in most of Latin America, these imports have done much to improve Costa Rica's cattle industry. During one nine-year-period (1969-1978) over 1800 registered Brahmans were imported by the Costa Ricans. Costa Rica has

*It is hard to overstate the importance of these visiting ABBA teams which often coordinate their visitations with the FAS of the USDA. They have been extremely effective in promoting good will, creating market interests and providing forums for the exchange of information. The ABBA has also helped to organize numerous Brahman clinics in conjunction with local associations in many parts of the world. (See Chapter VIII, "The ABBA Continues to Serve.")

Costa Rican commercial cattle of American Brahman blood are a common sight on Costa Rica's estancias.

remained a steady customer for the United States during the last decade, and this republic has also exported a number of Brahmans to Nicaragua during the past year. Another hopeful sign in Costa Rica for breeders of American Brahmans has been the demand for American Brahmans for the upgrading of dairy herds.[62]

Some of the prominent Brahman breeders in Costa Rica have been Oscar Pacheco, Arthur and Victor Wolfe, Jose Pinto, Jose Alberto Pacheco, Edgar Sanchez, Julio Tomeu, Alvaro Clachar, Tomas and Fernando Batalla, Victor and Julio Nino, and Jose Ramon Gonzalez.[63]

Panama, the first part of the continent to start raising any cattle, imported its first registered Brahmans in 1941.[64]

In 1942, when all shipping was a great risk because of Axis submarines, a Senor F. Arias P. of Panama purchased thirty head of registered Brahman yearlings from the J. D. Hudgins Ranch at Hungerford, Texas. Hauled by rail to California, they were loaded on a large steamer bound for Panama. As the ship was moving down the West Coast of Central America, it was engulfed by a huge fog bank causing it to run aground and wreck. The Brahmans were not injured, but they were all kept aboard while repairs were being made on the stranded ship. After a long wait, at which time the Brahmans were put on short rations, the ship was refloated and finally reached its Panamanian destination. While the Brahmans lost considerable weight during their ordeal, once they were pastured on the Arias ranch, they soon regained their lost poundage and contributed considerably to the upgrading of Senor Arias' fine herd.[65]

This ranch is now run by B. Guillermo St. Malo. Another great herd of American Brahmans in Panama is owned by the Motta Brothers.[66]

In September of 1980, a group of fifty-one self-described "small ranchers" from Panama made a whirlwind tour of Florida and ended up purchasing 200 American Brahmans and a number of registered quarter horses.

Organized by the Florida International Agricultural Trade Council, the Eastern Brahman Association and the Marketing Division of the Florida Department of Agriculture and Consumer Services, a tour for these Panamanians included visits to Brahman herds owned by the U. S. Sugar Corporation, the W. H. Stuart Ranch, the Chapman and Collier Ranch, G. T. Stack and Sons, Crescent O Ranch, G. A. Tucker and Sons, Cajuns Cattle Service, Heart-Bar-Ranch, Dr. Charles Reid and Florida Ranch Enterprises.[67]

Dr. John Stiles, Director of the Division of Marketing for the Florida Department of Agriculture and Consumer Services, reported that the Panamanians were very impressed with the Brahman cattle and the cattle operations they saw in Florida. "They also expressed optimism over the future for American Brahmans in Panama." An estimate of the total sales generated on the single tour was $600,000.[68]

The Inter-American Development Bank has approved a loan of $10.4 million, which with an additional $5.6 million loaned by the government of Panama, is expected to finance a $16 million

Bos indicus **cattle in India, such as these Gir cattle of Kathiawar, India are used principally as dairy cattle and draft animals.**

livestock development project in Panama to be completed by 1983. This progressive project will grant long term credit to about 350 cattle ranchers in Panama. Under this agreement the ranchers will be required to purchase supplies and equipment from Panamanian firms but will be free to purchase cattle abroad.[69]

One of the most active import markets of American Brahmans in Central America during the past decade has been the Republic of Guatemala. Since 1969, at least 3,622 registered American Brahmans have been brought into this land of purple mountain ranges and thick green jungles.[70]

Some of the first *Bos indicus* cattle to reach Central America were brought to Guatemala from the West Indies, and in 1905 a Brahman-type cattle herd was established from ten cattle purchased in England from the famous Hagenbeck Circus by Senors Luis Pedro Aguire and Carlos Hegel. They were landed at Puerto Barrios and trailed overland to the famous Hacienda Pachonte.[71]

Brahman cattle were also later brought into Guatemala by the United Fruit Company from 1921 to 1924. But it was not until 1945 that the first American Brahmans began to be imported. These were also brought in by the United Fruit Company. More *Bos indicus* cattle were also brought in that year from Jamaica by the sugar cane firm of El Saltos.

Through a series of political events, much of the United Fruit's cattle were sold to Manuel Ralda, then owner of Hacienda Pachonte. Part of these were American Brahmans.

The cattle operations of the United Fruit Company eventually came under the able direction of Neil Potter, ABBA Director, who bought both the herd and land holdings of the company. Over the years, Potter has been a driving force in the promotion of American Brahmans in Latin America.[72]

In Guatemala's mountainous neighbor to the east, Honduras, Criollo cattle are still common in the more mountainous regions, particularly in more remote farming areas in the East. But since the 1940's, many of the Hondurian cattle have been upgraded with American Brahmans. And today there are a number of excellent, purebred herds. One of the largest and best herds is owned by the famous Tela Railroad Company.

Another important Brahman breeder in Honduras has been Senora Gloria Flores de Losada of San Pedro Sula. The Hondurian government also maintains a sizeable herd at La Lima, Honduras.[73]

Honduras has not imported as many American Brahmans as most Central American countries; however, from 1969 to 1980, approximately 500 American Brahmans were brought in from the United States.[74]

Nicaragua first began importing American Brahmans in 1950, and a number of ABBA teams worked hard to promote the Brahman's general acceptance, particularly during the 1960's.

At one time Nicaragua had some of the best American Brahman herds in the world.*[75] Most of these cattle were evidently destroyed or exported during and after the Sandinista takeover, however, even under the present regime some attempt is apparently being made to rebuild Nicaragua's cattle industry. A government agency known as INRA recently purchased a quantity of Brahmans from Costa Rican breeders; however, their purchases were limited to $800 per head.[76]

Tiny El Salvador, which has the highest population density in Central America, has also been victimized by civil discord. But within an eight-year period, 1970-1978, this small republic had imported almost 1,100 head of registered American Brahmans. While it may be sometime before political and economic conditions are again stabilized in El Salvador there seems to be no doubt that more Brahmans will be needed there in the future for upgrading beef and dairy herds.[77] "The ABBA and the USDA/FAS were instrumental in helping a government program in El Salvador named MEGA get off the ground."[78]

American Brahmans in Africa

As we noted earlier, the influence of *Bos indicus* cattle has been apparent on the African continent for milleniums, but until the last few centuries - except for a few regions - cattle were not selected for economical reasons or functional efficiency. And it was not until the 1950's that registered American Brahmans were imported by any African nations, but generally where they

*Through a market development program initiated by the ABBA and the USDA's Foreign Agricultural Service, with the cooperation of INFONAC (Instituto de Formento Nacional), approximately 1,400 registered Brahman heifers were exported to Nicaragua from the United States with the purpose of setting up new breeding programs. During the 1960's, the ABBA set up an "Appendix Recording System," for Nicaragua by which special certificates were issued to Nicaraguan breeders trying to upgrade their herds but whose cattle were not yet 31/32 pure Brahman.(Harry Gayden, former ABBA Executive Secretary, Houston: Sept. 4, 1981.)

have been tried and given a fair chance, the results have been extremely gratifying.

The southern and eastern regions of Africa have thus far been the only parts of Africa that have imported or utilized American Brahmans in any significant quantities. But many areas of Africa still have the potential for developing excellent American Brahman and Brahman-derivative breed herds. Climatic and range conditions lend themselves well to increased Brahman production in such countries as Nigeria, Chad, the Congo, Ghana, Mauritania, Senegal and Angola.[79]

Some ranges, such as the northern savanna of Nigeria are already overstocked under present water conditions; however, beef production could be substantially increased with the proper management. In Ghana, for instance, by upgrading present herds with Brahman-blooded cattle and by the general improvement of existing management methods, enough beef could be produced to meet the country's own needs as well as producing enough surplus to export cattle.[80]

With proper management, the vast savannas south of the Congo's rain forests have a great potential for the production of American Brahmans and other *Bos indicus* breeds. Today large sections of these grasslands are unused except by wildlife. The Niari Valley, for instance, is covered with savanna vegetation. Since the 1960's, the government of Brazzaville has sponsored the raising of some cattle which are resistant to the dreaded tsetse fly. In any section of sub-Saharan Africa, the habitation of both livestock and humans may be restricted by this insect.[81]

Above the 10° North longitude in Chad, for instance, where there is no danger of sleeping sickness, there is extensive stock raising - occasionally in association with other agriculture. It is an area, however, that could still support much larger cattle operations with adequate stockwater. As in much of Africa, existing herds could be greatly improved with the infusion of high quality Brahman cattle. One of the greatest problems today that hinders increased beef production in this region is the isolation of ranges with a lack of adequate transportation facilities.

As is common in much of Africa, cattle in these regions generally are held by native owners for prestige, and numbers of cattle are the important thing. There is thus a reluctance to sell cattle for slaughter purposes or to even improve existing herds. With only elementary improvements in management practices with a greater offtake of existing cattle, beef production could be increased in many areas.[82]

It has been said that one of the world's best remaining cattle raising regions that is still greatly underutilized is an area that stretches across Africa from the Atlantic to the Indian Ocean between the 5th and 25th southern parallels. This area which is about 2,000,000 square miles in extent, includes the countries of Angola, Zambia, Mozambique, Botswana, Zimbabwe (formerly Rhodesia), and parts of the Congo, Tanzania, and South-West Africa. American Brahmans have been imported into parts of this area since the early 1950's, but they have been only utilized to any important degree in the eastern part of this region, particularly Zimbabwe.[83]

In 1954 a number of registered American Brahmans were purchased by the firm of Leibigs, Ltd., of Bulawayo, Zimbabwe. Placed on the company's huge ranch in the Matabeleland region, these Brahmans were crossed with existing herds of African and European stock. The results were remarkable with the average weight of mature Brahman-crosses running over 200 pounds more than that of the commercial cattle already there. Leibigs has since established large herds of purebred Brahmans and instead of growing out steers, they soon started selling heifers or "weaners" as they are called there.[84]

From 1972 to 1976, Zimbabwe - then Rhodesia - imported over 2,000 purebred American Brahmans. But as the scope of political and military unrest began to increase during the late 1970's, imports practically stopped - only eight Brahmans being shipped from the U.S.A. to Rhodesia between 1976 and 1977.[85]

A recent visit by a FAS/ABBA team to Zimbabwe indicated that potential marketing opportunities are good there and that the progress of Brahman herds in Zimbabwe and in nearby Zambia have been good. ABBA's Wendell Schronk, who was part of this team, pointed out that the need for U.S. Brahman seedstock is still apparent. At cattle sales he attended in Zimbabwe, he noted that those cattle with the most American Brahman breeding consistently brought higher prices.[86]

Cattle buyers from Zimbabwe were only recently in East Texas where they purchased Brahman herd sires from several ranchers.[87]

In many ways the American Brahman is ideally suited to the rangeland of Zimbabwe and this general region because of the Brahman's high tolerance to decimating cattle diseases common in this area. The climate is generally mild, and the stocking rate is approximately one cow for 15 to 25 acres. A Brahman breeder from Zimbabwe who was recently visiting the McKellar Ranch at Mt. Pleasant, Texas said, "there would not be enough cattle in Texas to stock Zimbabwe if the rangeland was fully utilized." Zimbabwe also has a Brahman Association, the Zimbabwe Brahman Association. Peter Grant of Broadwell Farms is the current president.[88]

Adjoining Zimbabwe to the north is the African Republic of Zambia. A little larger than the state of Texas, Zambia has long been noted for its production of copper. But it is said that the utilization ". . . of the veldt in this part of Africa would permit a severalfold increase in the present cattle population of approximately 1,300,000." Presently this area, which lies on a broad plateau of about 4,000 feet above sea level, has one of the lowest cattle densities in Africa "with only about 4.4 head to the square mile."[89]

Commercial cattle breeders in Zambia have purchased only a few purebred Brahmans from the U.S.A. Most of their purebred Brahmans have come from South Africa, but the FAS/ABBA team that visited Zambia in the summer of 1980 found that Zambian buyers could actually purchase cattle and import them cheaper from the United States. "Several (breeders) were working on a plane load of U.S. Brahmans in early 1981 to be shipped to Zambia."[90]

According to ABBA's Wendell Schronk, "One of the biggest problems with foreign trade to these two countries (Zimbabwe and Zambia) is the exchange of foreign currency. Cattlemen say it takes six months to arrange for this currency exchange. If the U.S. government will help to insure the funds or will assist in this problem, there would be a tremendous market open for our breeding cattle."[92]

Zimbabwe's neighbor to the east, Mozambique, also lies in this region between the 5th and 25th parallels and probably has the potential for increasing its cattle population some tenfold. The first American Brahmans were brought to this beautiful tropical nation in 1955, and some imports were made on a rather regular basis until 1972. Since then no transfers have been recorded by the ABBA. A few Brahman-derivative breeds were also brought in during the past decades. While the American Brahman appears to be a natural for increasing beef production in this area, their impact there has thus far been inconsequential.[93]

Within the broad belt of land between the 25th and 5th parallels in Africa are also the countries of Botswana and South-West Africa or Namibia. While both areas are very dry, they cover an area of over 550,000 square miles. Cattle raising is the chief industry of Botswana although the national cattle population is only 1,347,000. Most of the cattle are kept at so called *cattle posts* and are herded by young boys. American Brahmans were first brought into this area in 1964 but have not yet been utilized to any degree as there has been no concerted effort to upgrade existing herds.[94]

In South-West Africa most commercial cattle raising is carried on in Northern Namaqualand around Windhoek, the capital. The first American Brahmans were brought to South-West Africa in 1953 by a European breeder named J. R. Cranz. Cranz was in South Texas during his honeymoon, and he and his wife were so impressed with the gentle Brahmans that they purchased 17 head from the J. D. Hudgins Ranch. Shipped by steamer from Galveston, the cattle were at sea for 26 days before being unloaded at Cape Town, South Africa. From Cape Town they were loaded on freight cars and shipped some 800 miles to

Windhoek where they were unloaded and trailed to the Cranz ranch. All seventeen Brahmans made the trip in good shape becoming the nucleus of the Cranz Brahman herd.[95]

The importations of American Brahmans into South-West Africa have remained fairly small and irregular. The largest importation was made in 1975 and 1976 when 68 registered Brahmans were brought from the United States. The Africander breed is still the most important breed in South-West Africa, and the chance of expanding beef production may be limited at the time, although large areas of dense acacia with low forage production offer some possibility for expanding beef production. The stocking rate is approximately one animal for 25 acres. Hopefully the rustling ability and good beef qualities of the American Brahman will produce a greater desire for more importations in the future.[96]

By far, the African nation that has done the most to establish the American Brahman is the Republic of South Africa. Occupying the southern end of the great African continent between the 22nd and 35th southern parallels, this nation provided breeders of American Brahmans with one of the best markets in the world from 1954 to 1976. During one two-year period (1974-1975), 2,284 head of registered Brahmans and 2,637 head of Brahman-derivative cattle were exported from the United States to South Africa.[97] And today it is estimated that there are approximately 65,000 registered Brahmans in this hospitable land of sweeping veld, mountain ranges and narrow coastal plain.[98]

In their attempt to establish high quality Brahman herds the South African Brahman breeders have seemingly spared nothing as far as expense goes. According to a leading Brahman breeder and exporter, G. T. Stack of Tampa, Florida, "Money never seems to be a problem. What the South African breeders want are the top of the tops, and they are getting them with the American Brahmans." While imports of American Brahmans have dramatically dropped since 1976, some still continue to be imported from the United States. In 1980, the ABBA recorded the transfer of only 14 head of registered American Brahmans to

South Africa, but hopefully this amount will increase in the years ahead.[99]

The competitive quality of Brahman cattle in South Africa and the influence that American Brahman bloodlines have had on South African herds was recently demonstrated at "Phase A" of the South African National Beef Cattle Performance Testing Scheme. At this show Bob Tecklenburg of Barberton, South Africa showed a registered Brahman cow, *Tecklenburg Miss Resoto 45*, that topped 82,000 other cows of all breeds. She was descended from the bull *Resoto Manso* who was the champion bull at the 1942 Houston Fat Stock Show.

It was noted that *Miss Resoto 45* had an average intercalving period of 369 days and had had eight successive calves with an average weaning index of 110 - an extremely impressive figure.[100]

Today South Africa has a very active Brahman breeders association, the South African Brahman Breeders Association, which has been headed by J. Alf Coetzer for about 15 years. Currently the Executive Secretary is Dave Morley. The SABBA is one of the strictest breeder associations in the world regarding quality control. One of its policies over the years has been to make periodical blood type checks on sires and dams being proposed for registration.[101]

Today South Africa has one of the most progressive cattle industries in the world. While this progress is especially evident among so-called European breeders, the Bantu, or native herdsmen also have better cattle than are found in most parts of Africa. Despite this, there is still considerable room for expanding present production.[102]

While the Africander and the Bonsmara breeds will continue to compete with the American Brahman in the coming years, the American Brahman has firmly established itself as a desirable breed since it was introduced less than thirty years ago. And its popularity will undoubtedly remain strong in this African republic.

American Brahmans in Asia

While *Bos indicus* breeds have been a part of the Asian landscape for milleniums, until recently few had been used for commercial cattle operations. In countries where there is still little mechanized farming and frequent food shortages cattle have obviously been used as draft animals and, or as, milk animals.[103] To most farmers in Southeast Asia, the slaughtering of one's cattle would be tantamount to setting fire to one's tractor or burning one's food. However, during the last thirty years or so, a growing number of Asian governments have begun to promote commercial cattle operations. And a number of Asian livestock scientists and breeders have turned to the United States and other nations, such as Australia to find superior cattle for developing beef herds.[104]

A high Indian official in the Ministry of Agriculture visited the United States a few years ago and observed, "One must marvel at the grandeur of the American Brahman when compared with its parental types in India. We have nothing like it in our country."[105]

But the first Asian nation to actually begin importing registered American Brahmans was the Republic of the Philippines, beginning in 1952. This country, consisting of a group of some 7,000 beautiful emerald islands, has always had the basic elements for a healthy cattle industry.[106] Much of the land is unsuitable for cropping, but it has had the right soil types and the abundant rainfall that can even now produce fodder for several times the present cattle population.[107]

Since 1952, there has been a steady, though sometimes small, exportation of American Brahmans from the United States to the Philippines. From 1970 to 1979, appoximately 600 head of registered American Brahmans left the United States for the Philippines. Brahmans from Australia have also been shipped to these islands over the years.

The leading breeder of American Brahmans in the Philippines today is Antonio Nocom of Manila, who runs about 500 purebred Brahmans on his beautiful Mindanao Island ranch. The

Valley Dairy, located some twenty-five miles from the city of Davao on Mindanao, has a joint dairy and beef herd which includes a number of registered American Brahmans and Santa Gertrudis bulls. Progressive breeders like Nocom have been big boosters of registering their Brahmans with the ABBA, and between 1974 and 1978, Philippine breeders had registered over 1,100 Brahmans with the ABBA.[108]

According to the Foreign Agricultural Service of the USDA, exports to the Philippines will be small during the next few years. However, if the government's ten-year beef-carabeef (buffalo) program is carried out, commercial ranchers probably will import to introduce new bloodlines and to upgrade present herds. "The main problem today is one of management (nutrition and disease control) of newly arrived animals for at least the first 6 to 8 months or until imported animals become acclimated."[109] Brahmans and Brahman-crosses are, of course, better able to make this transition to a tropical climate.

Philippine government officials are reportedly sold on the American Brahman. But at the present time, all importing costs must be taken on by the individual breeders, and most are not able or willing to make such outlays of capital.[110]

Two years after the Philippines began importing American Brahmans, the ancient country of Siam - now Thailand - imported its first American Brahmans. As most of the Thai people are Buddhist, there has never been a need for any substantial cattle - beef operation; however, because of the large number of European hotels and Western visitors in Thailand, some years ago the Thai government started encouraging commercial cattle production. From 1974 to 1975, Thailand imported 778 Brahmans from the United States. A Western type abattoir was built some years ago with U.S. AID money at Bangkok, and the Thai government has since established a sizeable Brahman herd.[111]

R.W. Mayronne, Jr. of Covington, La., has arranged most of the recent American sales and transfers of registered American Brahmans to Thailand. Most have been bulls. While the greatest number of these cattle have gone into government progams, a

A herd of *Bos indicus* cattle graze on Sumba Island, Indonesia with Arabian horses. In the late 1940's there were over 22,000 Indian cattle on this island.

number actually have been dispersed for breeding purposes to farmers in the highlands. Importantly, the government now recognizes the need for high quality breeding stock, and hopefully a greater demand will arise for the importation of American Brahmans.[112]

The Federation of Malaysia is another Southeast Asian country that has started importing American Brahmans. Since 1971 approximately 150 Brahmans were brought in from the U.S.

The Malaysians have struggled for a number of years to develop a better dairy type, but have had little success. A rather obscure mixture of various types of Zebu cattle from India, referred to simply as the L.I.D. cows, have been used for draft and dairy purposes. However, the Malaysians are also currently crossing F_1 Hereford X Brahmans with Holsteins which show signs of producing a superior dairy cow.[113]

Between 1973 and 1975, 4,599 F_1 Hereford X Brahman were exported from the United States and Australia to Malaysia.

A number of registered Braford cattle were also exported to Malaysia during this same period.[114]

USDA reports indicate that there will be a sizeable importation of breeding cattle into Malaysia over the next few years (up to 15,000 head per year according to one report).[115]

Mr. Byron Johnson, ABBA Director, reported after a trip to Malaysia in 1977 that the Malaysians had embarked on a very comprehensive cattle program, and during 1978 the Malaysian government contracted with the Australians for the sale of 10,000 head of Droughtmasters to continue building up their dairy herds.*[116]

It would appear that most of the Brahman-crosses going to Malaysia and Thailand in the near future at least will come directly from Australia because of Australia's traditional ties with the Malaysians and Malyasia's dependency on the World Bank for capital.[117]

Beginning in 1957, the Republic of South Korea imported the first American Brahmans.[118] While some American Brahmans and Santa Gertrudis bulls have been crossed with native cows, nothing of great significance has developed from this early experimental work. Korea's limited pasture land and the chronic negative balance of payment problems have caused the South Korean Government to closely monitor the importing of livestock. However, a sizeable number of Brangus, Holstein and Angus cattle have been imported since the early 1970's.[119]

The largest beef operation today is on the Island of Cheju in the East China Sea. It is the 2,000 acre Samho Cheju Ranch.

*The Droughtmaster is a cross developed from the American Brahman on the Shorthorn. This cross is found in abundance in northern Queensland and also is raised on other parts of the Australian Continent. The use of Brahman bulls in Queensland herds has markedly increased during recent years. (See Chapter VI, "The Crossbreeding's Common Denominator," and Wendell Schronk's, "Brahmans from the Melbourne Zoo," *The Brahman Journal*, Feb.,1980.)

Here American Brahman and Santa Gertrudis bulls are being crossed with indigenous Korean cattle and so-called Korean Blacks.[120]

Since the end of World War II, there has been a general increase in the world trade of beef and dairy cattle. While the United States has by no means been the only cattle exporter -nor indeed the largest, it has certainly provided the world market with some of the best quality breeding cattle. And the American Brahman and American Brahman-derivative breeds have been the number one export breeds. Besides helping the U.S. trade balance, and thus helping in a positive way to stabilize the American dollar, many of these cattle have aided significantly in improving the quality of animal herds around the world - especially in Latin America, Australia and South Africa.

While the largest number of American Brahman sales will continue to be domestic in the coming years, historically the exportation of American Brahmans has certainly been important to the cattle industry. Some breeders, indeed, have concentrated on foreign sales. One of the largest breeders in the USA recently said that foreign transfers over the past 45 years averaged 41 percent of his sales with some years over 85 percent.[121]

Laws relating to the importation of livestock change over the years in various countries, and sometimes current regulations appear to be completely unrealistic. Permits for importing cattle into Venezuela, for instance, are almost impossible to obtain today, whereas five years ago (1976) over 1,500 American Brahmans were exported from the USA to that country. Ecuador, which has been a good market for purebred American Brahmans, has a very complicated set of rules regarding pedigrees and production tests, particularly for dairy cattle. However, market conditions may soon improve in both of these countries for the importation of American Brahmans.[122]

Australia suspended the importation of American cattle some years ago. Edgar Hudgins recalls that in 1954 a shipload of American Brahmans and Santa Gertrudis was destined for Queensland when the Australian government decided to suspend further imports. "Of course," recalled Hudgins, "the

Edgar Hudgins, patron of the Australian Brahman Breeders Association, and son Leslie Hudgins, helped judged these two and three year old registered Brahmans at the Brisbane, Australia Royal Show in 1971.

Aussies honored our contracts, and we were able to unload our cattle, but that was the last load of American Brahmans that has gone there since." Needs and laws change, however, and the Australians are scheduled to start issuing new permits for the importation of cattle for 1982.* [123]

*Approximately 30 to 50 head of American Brahmans are scheduled to go to Australia this year (1981). They will first be transported to Balzac, Canada (near Calgary, Alberta) for a whole series of tests before being shipped to Australia.

Of course, one of the real keys to good foreign markets today for Brahmans is political stability in potential markets. In those countries where severe internal conflicts prevail or generally where Marxist regimes have gained control, there has been no demand for the importation of purebred American cattle of any kind. The last shipment of American Brahmans to Nicaragua, for instance, which was once a good export market, was in 1979.[124]

Notwithstanding the existence of some problems in the exportation of Brahmans and other American cattle, Dr. James R. Simpson, Marketing Economist from the University of Florida, points out that the limits placed on cattle imports under the Meat Import Act of 1979, the emphasis on expanding beef-breeding cattle exports by the USDA and the increased activity of such organizations as the Florida International Agri-business Trade Council will increase future foreign demands for American cattle.[125]

A Breed for the Present and the Future

Speaking in 1947 at the first American Brahman Congress, the Honorable Judge Frank Scofield said, "There are tremendous possibilities in the Brahman business, the greatest future today that any breed of cattle on earth ever had. In the heyday of the Hereford, the Angus, the Shorthorn, they never started to have the possibilities that the Brahman breeders of the world have today. . . ."[126]

Spoken over thirty-four years ago, some of these possibilities have been realized. On the spacious Llanos Plains, the Gran Chaco and the Pampa of South America, on the Great Karro of South Africa, on the broad Coastal Plains of North America and in many other native ranges around the world, the American Brahman and its crosses have become a permanent part of the landscape. In habitats once considered unfit for cattle, the

Brahman and its crosses are surviving and usually thriving. And with the American Brahman, ". . . generally recognized to be the most highly developed beef breed of the *Bos indicus* species . . ." the best may be yet to come.[127]

Many agricultural barometers explicitly show that there will be a greater demand for beef and dairy production in the world in the years ahead. Much of the unused and poorly utilized rangelands where cattle production can be increased are best suited for Brahman-blooded cattle. Indeed, in many areas they are the only cattle that can subsist.

At the peak of America's war effort in 1944 when certain foods and fuels were being rationed in the United States, J. W. Sartwelle said in an address to the ABBA Directors, ". . . the Brahman breed can do better on the feeds that we have, and we should capitalize on this ability of less expensive feeds."[128] While Sartwelle's remarks were prompted by a sense of patriotism during World War II, they are no less true today.

Dr. Stewart Fowler wrote in 1974:

> *Due to the tremendous upsurge in the world's population, we may be approaching the day when we can no longer afford the luxury of channeling 6 to 8 pounds of grain through the ruminant digestive system to be converted into a pound of gain which will yield less than a half pound of America's favorite food. . . We could find one of these days that we must utilize our corn, milo, barley and other feed grains directly through the stomach of man in order to utilize every pound most efficiently . . . This means that we would need to make fuller use of the fermentation vat of the bovine compound . . . to utilize ever-increasing amounts of roughage and other materials not readily utilized by man.*[129]

It is the *Bos indicus* breed, its derivative breeds and its crosses that can survive and produce in restrictive environments, and that can thrive in normal environments. As we have seen, this can be attributed to the breed's genetic ability to utilize forages more and to graze during the hottest part of the day.

Many Brahman breeders today in many parts of the world have taken advantage of the unique qualities of the Brahman and market-feed cattle out on grass. Rancher and ABBA Director Gilbert Tucker of Cocoa, Florida, is one of many breeders who has established such a program. His commercial cattle are Brahman X Angus which go from birth to market on grass. "Improved pastures feature White Dutch Clover, St. Augustine, Pangoa, Argentine Bahia and Big Stemmed Hermathria . . . . . Steers are kept until they are two years old and then sold for slaughter . . . they will weigh 1,100 pounds and grade standard and good."[130]

In Guatemala, on the ranch of Luis Morales, range-raised Brahman cattle are grown out on African Star grass and are fattened on Pangola. From the Pangola pastures, Morales' Brahmans and Brahman-crosses go directly to the packing plant.[131]

Sometimes the advantages of Brahman-blooded cattle are overlooked in more affluent societies. But an economic adaptation trait that has become increasingly important is the fact that the Brahman hybrid carcass normally contains more beef and has less fat. Because of this factor, it has won the approval of doctors concerned with the heavy consumption of animal fat by human beings.[132] And for those retailers and consumers who want the highest possible portion of edible, tasty meat with high cutability, the Brahman type is the answer.[133]

The recognition of the Brahman as a breed having the ability to produce the lean, choice quality beef desired today was emphatically shown in 1979 when *R.Q's Rexcrata 455* of the Diamond T Ranch was designated as a "Certified Meat Sire." A highly respected accolade in international breeder circles, "The Certified Meat Sire" is the branch of Performance Registry International for identifying and recognizing a seedstock sire who is capable of producing the beef traits desired today. Animals competing for this honor are actually fed under feedlot management techniques, but they are fed to an estimated low choice grade in order to avoid more than 4/10 inch outside fat." It is only after a period of meticulous evaluation that a sire is so selected.[134]

Disease and insect resistance, heat tolerance, longevity, good maternal instincts, outstanding rustling ability, high feed utilization, hybrid vigor, lean meat, lifetime fertility, natural selection, extraordinary adaptibility and versatility and excellent marketability are all epithets that we have used in reference to the American Brahman. But these words and phrases are much more than just promotional catchwords. Years of extensive research, accumulated data and breeding experience have revealed how well they truly describe the Brahman.

Surely no other breed in the world has expanded its influence so extensively and so rapidly as the American Brahman. If there is indeed a single total breed that is a "breed for all seasons," it is the American Brahman.

John Joyce of Wharton, Texas (co-owner of the Seaberg Ranch), is the originator of the famous Brahman clinics, the custom fitting service for Brahman cattle and the first sales management firm for purebred Brahman sales. He was principally responsible for the first sale of a $100,000 Brahman bull when *R Q's Rexcrata 455* was sold by 3W Ranches of Donnie, Texas to the Diamond T Ranch of Bay City, Texas. Besides organizing Brahman clinics in the U.S.A., he has also put them on in Mexico, Central America, South Africa and Zimbabwe (formerly Rhodesia).

Courtesy Diamond T Ranch

In 1979 *R Q's Rexcrata 455*, of the Diamond T Ranch was designated as a "Certified Meat Sire." It is only after meticulous evaluation that a sire is so acclaimed, and this accolade is highly respected in international breeder circles.

Left - Henry Koontz, prominent Brahman breeder in the 1940's and early 1950's, was owner of the Koontz Ranch at Inez, Texas. "Mr. Henry" was an active member of the ABBA, serving as a director from 1950 to 1954. During the 1940's, Koontz exhibited his Brahmans at the major show circuits in North America and Latin America.

Below - one of the early Brahman breeders of Alabama was Wilson Coker pictured here at his farm near Hayneville, Alabama with two young American Brahmans. Coker bought his early breeding stock from the Norris Cattle Company of Ocala, Florida and from M. C. Stallworth of Hope Hall, Alabama. This picture was made in 1948.

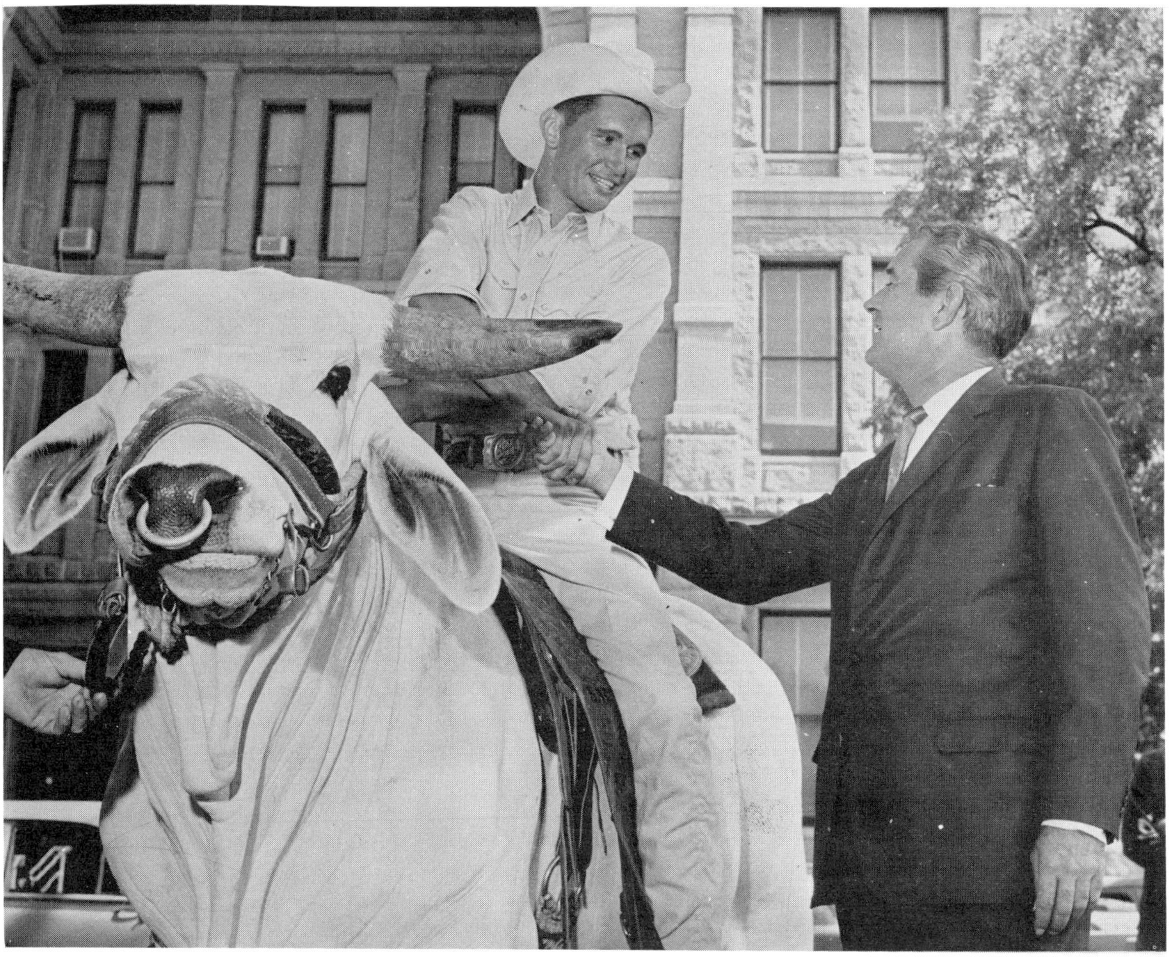

In 1964 Governor John Connally (R) proclaimed this Brahman bull named *Bobo* as "a Texan Ambassador of Goodwill." Shortly after this picture was made, Texan Jerry D. Cotten (astride bull), rode *Bobo* from Saginaw, Texas to the World's Fair being held in New York City. Cotten's trip was over 2,000 miles, and it took approximately seven months for him to complete it.

Bibliography

Sources Consulted

Adams, Alto, Jr. "Braford Breed Originated in Florida." *Florida Cattleman*. Florida Cattlemen's Assoc. Kissimmee, Fla: Jan. 1976.

Akerman, Joe A., Jr. "Rustling, Dipping, Fencing and the End of an Era." *Florida Cowman*. Florida Cattlemen's Assoc., Kissimmee, Fla.: 1976.

ABBA Bibliography of Articles, Books and Other Publications on Brahman Cattle. Property of ABBA. Houston, Texas. Some of the publications go back to early 1900's.

American Brahman Breeders Association's 50th Anniversary Banquet. (Tape), property of ABBA. Houston, Texas, Feb. 21, 1974.

"American Brahmans - the Great Southern Beef Hybridizers." Progressive Farmer, March, 1974.

American Brahman Breeders Association Minutes, 1924-1980. ABBA office. Houston, Texas.

"ABBA Registrations and Transfers, 1924-1973," *The Brahman Journal*. Chaparral. Sealy, Texas, Feb., 1974.

"ABBA Registrations and Transfers, 1939-1980," ABBA office. Houston: 1980.

"Ambler's Brahmins in 1858." Letter from Dr. A. C. Ambler to Florida State Senator Phillip Dell regarding use of Brahmans in Florida in 1858. Cited from *Florida Cowman* by Joe A. Akerman, Jr. Florida Cattlemen's Assoc., Kissimmee, Fla., 1976.

Angus-Brahman F_1 Field Day. Brooksville, Fla. July 19, 1980.

"Australian Market to Reopen." *Zebu Journal*. Cotulla, Texas. July-Aug., 1980.

"Australia To Import American Beef." *Zebu Journal*. Cotulla, Texas. July-Aug., 1980.

"Baker Herd was Divided Between Son and Daughter." *International Brahman Review*.

Barton, Catherine. "Visiting the 3X Brahman Ranch." Houston: *Brahman Breeder-Feeder*, May, 1946.

Bassford, Forrest. "The Indomitable Bos Indicus," *Western Livestock Journal*, Sept., 1970.

"Battle Island Ranch." *Red Brahman Directory*. Property ABBA, Houston, Texas.

Baughman, J.L. "Climate, Cattle and Crossbreeding," *Texas Journal of Science*, Vol. III, No. 2, 1951.

"Better Bulls Get More Pounds and Uniformity at G Bar E Ranch." The *Florida Cattleman*. Florida Cattlemen's Assoc., Kissimmee, Fla., July, 1978.

Black, W.H., Senior husbandman. Bureau of Animal Industry. USDA. "Brahman Cattle in America," *American Feed and Cattle Dealer*. (Reprint), 1947.

Black, W.H. "Development of Guzerat-Aberdeen Angus Crossbred Cattle." *Coastal Cattleman*, Beaumont, Texas. March, 1937.

Beef Cattle in Florida. Bulletin No. 28. Fla. Dept. of Agriculture and Consumer Services, Institute of Food and Agricultural Sciences, Gainesville. 1976.

Blassingame, Wyatt. "Mr. Henry's Empire." *Saturday Evening Post*, Jan. 7, 1956.

Bluebonnet Farms Research Studies 1948-1953. McGregor, Texas. "Feeding Brahman and Brahman Crossbred Steers," Lloyd Clyburn, ABBA. ABBA Papers, Houston, Texas. April 25, 1955.

Borden, A.P. Executor of the Estate of A.H. Pierce. Letter to Hon. R.F. Broussard regarding Brahman cattle imported from India in 1906. Pierce, Texas. May 29, 1911.

Borden, A.P. "Report to American Breeders Society, 1908." (Copy), ABBA Papers, Houston, Texas.

Bray, C.I. La. Agricultural Experimental Station, "L.S.U. Professor Tells of Experimental Work Done With Crossbreds," *The Florida Cattleman*. Feb.,1952.

"Eola Border Dies Tuesday at 100." The *Wharton Journal-Spectator*. Wharton, Texas. April 3, 1980.

Bracher, Bob. "Brahman Bulls on Angus Cows." *Cattleman Magazine*, Fort Worth, Texas. 1972.

"Braford Cattle Rapidly Coming to Forefront in Coastal Livestock Picture." *Coastal Cattleman*, Beaumont, Texas. June, 1940.

"Brahman-Angus Cross Proving of Value in South." USDA Office of Information, Nov. 30, 1934.

Brahmans Australian History. Australian Brahman Breeder's Assoc., Rockhampton, Qld., Australia, 1972.

"Brahman Breeders Organize First Association in Ohio." *Springfield Daily News*, April 12, 1952.

"Brahman (Zebu) Cattle," *Farmer's Bulletin*, 1361. USDA. Washington, July, 1924.
"Brahman Cattle are 'Getting Around' in the Far West." *Brahman Breeder-Feeder*, July, 1946.
"Brahman F_1 To Be Association Certified." The *Brahman Journal. Chaparaal. Sealy, Texas, 1979.*
"Brahmans, Fad, Fancy or Fact?" Breeder Gazette, Houston, Texas, April, 1952.
The Brahman Handbook. ABBA, Houston, Texas.
"Brahman-Hereford Field Day, A First for the Industry." Sagebrush. Sealy, Texas, *Brahman Journal*, April, 1980.
The Brahman Hybrid. ABBA. Houston, Texas.
"Brahman Publicity on 1924 Importation." *American Brahman Journal*. Silver Anniversary issue. Houston, Texas, Feb., 1949.
"Brahman Registration Transfers, 1939-1978." ABBA. Houston, Texas,1980.
"Brahman Studies, 1920-1928 at King Ranch." *Breeder-Feeder*, Nov., 1949.
Bray, C. I. Animal husbandman, Louisiana Agricultural Experiment Station."Brahman Crossbreeding in Louisiana." *Brahman Journal*, Houston, Texas, Feb., 1949.
Bray, Charles. "Brahmans in Louisiana." *Gulf Coast Cattleman*, Nov., 1948.
Breed Crosses with Beef Cattle in the South. Southern Cooperative Series Bulletin 81, Mar., 1962.
"A Brief History of Brahman Cattle in the United States." *American Brahman Journal*, Houston, April, 1946.
Briggs. *The Brahman and Associated Breeds* (Property of ABBA). Houston, Texas, 1980.
Brockett, Joe Ed. "The American Brahman." *The Brahman Journal*. Sagebrush. Sealy, Texas, May, 1980.
Brockett, Joe E. "The Brahman Cattle of Hacienda Pachonte." *The Brahman Journal*. Sagebrush Pub. Co., Sealy, Texas, July, 1980.
Brockett, Joe E. "Family's Brahman Story." (Short history of Hudgins family), *The Brahman Journal*, August, 1978.
Brockett, J.E. "First Brahman Certified Meat Sire." *The Brahman Journal*. Chaparral. Sealy, Texas, May 1, 1979.
Brockett, Joe E. "From Out of the Swamp." *Brahman Journal*. Chaparral. Sealy, Texas, Feb., 1980.
Broyles, William, Jr. "The Last Empire," *Texas Monthly*, Oct., 1980.
Brockett, Joe E. "Homer Lord - Early Brahman Breeder." *Brahman Journal. Chaparral Press. Sealy, Texas, 1978.*
Brockett, Joe E. "A Look at Guatamala's Next Generation, Luis Morales." Brahman Journal. Sagebrush. Sealy, Texas, April, 1980.

Brooks, Harold J. Ph.D. "The Brahman Contribution to Commercial Dairying," *Proceedings of the First Brahman Congress.* Southeastern Brahman Breeders Assoc., Sarasota: 1947.

Bundy, Clarence, Diggins, Ronald and Christensen, Virgil. *Livestock and Poultry Production,* 4th ed. Prentice-Hall, Inc. Englewood Cliffs, 1975.

Butler, Milby. Letter to Mrs. Thomas about Brahmans, Aug. 20, 1945. ABBA Papers.

Carpenter, Z.L. "Effects of Crossbreeding on Carcass Characteristics." University of Fla. Press, Gainesville, 1973.

Cartwright, T.C. and Fitzhugh, H.A., Jr. Texas A&M Livestock Geneticists. "Crossbred Bulls?" *Crossbreeding, Tool for Profit or Financial Disaster?* American Angus Assoc., St. Joseph, Mo.

Cartwright, T.C. "Longevity," *Cattleman,* February, 1971.

Cartwright, Tom C. and Fitzhugh, Henry, Jr. "Selecting and Breeding the Unique Brahman," *Brahman Journal,* Sept., Oct., Feb., 1972-73.

"Cattle Kingdom of Kissimmee." *Tractor Farming,* Vol. 35, July-Aug., 1952.

Chambers, Lionel. "Central Romana," *Brahman Journal.* Chaparral Publishing Co., Inc., Sealy, Texas, Feb., 1974.

Clark, Mary Whatley. *A Century of Cow Business.* Texas and Southwestern Cattle Raisers Assoc., Fort Worth, Texas, 1976.

Clyburn, Lloyd. Director Information and Education, ABBA. Unpublished manuscript on American Brahman. Houston, Texas, April, 1955.

Colbert, Walter. "What It Is Now and What It Was 65 to 70 Years Ago." Early history of Texas fever tick. Property of ABBA, Houston, Texas.

"Consumer and Market Research on Beef," *Beef Business Bulletin,* NCA, Jan. 16, 1981.

Cope, Channing. "Robust Cattle Raised by Brahman Breeders." *The Atlanta Constitution,* Oct. 3, 1949.

Cornelius, Paul. Brahman breeder and meat packer. Letter to Avon Davis of Chowchilla, Calif., Sept. 17, 1973.

Cunningham, Kirby. Dir. Public Relations, ABBA. "Brahmans - Past, Present and Future." *Gulf Coast Cattleman.* Beaumont, Jan., 1971.

Cunningham, Kirby. "A Historic Herd for the Future." *The Cattleman.* Feb. 1971.

Cunningham, Kirby. Dir. of Public Relations, ABBA. Letter to Mrs. D.M. Parmelee of Austin, Texas regarding Mr. J.A. McFaddin. Houston, May 28, 1970.

Cunningham, Kirby. ABBA Ex. Secretary, "Fertility in the Brahman Cow," *The Cattleman,* Feb., 1971.

Cunningham, Kirby. Exec. Secretary, ABBA. "The Sartwelle Brothers, Texas Ranching Pioneers." *The Cattleman,* Feb. 1971.

Cuthbirth, Sam T. Gen. Mgr. Pierce Ranch, Limited. Letter to Dr. Burch H. Scheider regarding 1906 Borden-O'Connor Importation. Pierce, Texas, Sept. 13, 1949.
Dial, Joe. Brahman breeder and ABBA Director. "History of Brahman Cattle Registered in ABBA," Paper. Placedo, Texas, 1979.
Dobie, Frank J. *The Longhorns.* Little Brown, Boston. 1941.
Doris, Don. "Matching Cattle Characteristics to Climate Conditions." *Livestock*, Nov., 1979.
DuPont, William, Jr. Notes by Edgar Hudgins regarding William DuPont's development of a Santa Gertrudis herd in 1943.
Eddy, Don. "Cowhands on the Grapefruit Range." *The American Magazine*, Dec., 1949.
Ehmke, Vance. "Brahman Crosses For the Great Plains." *The Brahman Journal.* Sagebrush Pub., Sealy, Texas, Jan., 1980.
"Eighteen Imported Bulls," *Brahman Breeder-Feeder.* Houston, Texas, July, 1946.
Emmett, Chris. *Shanghai Pierce, A Fair Likeness.* Norman: University of Oklahoma Press, 1953.
"Emperor Leased to Texan." *The Stockman.* March, 1950.
Empty Tea Cup Fills Texas with Brahmans. Unpublished manuscript owned by ABBA. Houston, Texas, 1936-37.
"Epic Breed." Notes by Charles R. Koch on the American Brahman.
"Esteban and Eligio Garcia of McAllen, Texas." *Brahman Breeder-Feeder. Houston. Texas, May, 1946.*
"The Expansion of 100% Zebu Blood." *The Zebu Journal.* Cotulla, Texas, Sept., 1949.
"Expica '80." *The Brahman Journal.* Sagebrush Publishing Co. Sealy, Texas. July, 1980.
"Export Veteran Knows His Cattle." *Florida Market Bulletin.* Florida Dept. of Agriculture and Consumer Services. Tallahasee, Fla., June 15, 1977.
"Family Ranch Big Florida Business." *Toledo Blade*, Jan. 17, 1956.
"The Famous Manso." J. D. Hudgins, Inc. Hungerford, Texas, 1980.
"Famous Manso Discoverd by Hudgins in '31." *International Brahman Review*, XVII, Jan., 1961.
Finch, F.H. "Brahman-Guzerat Cattle and the Necessity of Crossing With Them in Northern Argentina." Estancia Garruchos, Provincia de Corrientes, Feb. 5, 1941.
"First American Brahman Congress (Proceedings)." Southeastern Brahman Breeders Assoc. Kissimmee, Fla., Oct. 30-31, 1947.
"First Brahman Field Day." *Brahman Breeder-Feeder.* Houston, Texas. Aug., 1946.
Fisher, Wayne H. "Brahmans in Hawaii." *The Zebu Journal*, Sept., 1950.

Fisher, Wayne. "United Nations in Agreement on One Subject - Livestock." Review of United Nations Publication, *Breeding Livestock Adapted to Unfavorable Environments.*, 1948.

Florida Department of Agriculture and Consumer Services. News Release, Tallahassee, Fla., Sept. 11, 1980.

Fowler, Stephen. *Beef Production in the South.* The Interstate, Danville, Ill., 1969.

Fowler, Stewart H. "The Original Exotic and Beef Revolution in the South." *Cattleman Magazine.* Fort Worth, Texas. Feb., 1973.

Francis, John. Prof. of Preventive Medicine and Anatomy, Veterinary School, Univ. of Queensland, Brisbane, Australia. *Definition and Use of Zebu, Brahman or Bos indicus Cattle.* Reprinted from *Nature*, Vol. 27, No. 4992, July 3, 1965.

Francis, Ruth Garrison. "Brahman Cattle Came Into Texas From Louisiana." *Coastal Cattleman*, Jan., 1942.

French, M.H. "The Failure of Pure and High-Grade European Cattle in Hot Climates." *The East African Agricultural Journal.* April, 1941.

"From History - Sam Houston - ABBA No. 1." *American Brahman Journal.* Houston, Texas. Feb., 1949.

Ganado Brahman Americano. ABBA Publication, Houston, Texas. 1974.

Gates, J.V., Brahman breeder. *History of Barzee Brahmans.* Poteet, Texas. 1978.

Gatewood, Boyd. "Cattle King's Domain Grows Richer Under Modern Management." *Houston Post Dispatch*, Nov. 22, 1925.

"The Gir Forest Ecosystem." A film on the Gir Forest of India and the affects of cattle grazing on this region. North Florida Junior College Library, Madison, Fla.

Guide for U.S. Cattle for Exporters. Agricultural Handbook, No. 217. USDA, Washington, 1976.

Halff, Mayer L. New York lawyer and former resident of San Antonio, Texas. Reprint of letter appearing in *Cattleman*, Nov., 1950, addressed to Chris Emmett regarding Halff's role in having President Theodore Roosevelt help save *Bos indicus* cattle of 1906 Importation. New York, Dec. 2, 1941.

"Hereford-Brahman Field Day, An Industry Pacesetter." *Weekly Livestock Reporter*, June 5, 1980.

"History and Definition of Brahman." 1964 Symposium. Property of ABBA, Houston, Texas.

"History of Herd, Cornelius Cattle Co." *Brahman Breeder-Feeder, Houston, Texas. Jan., 1946.*

"*The History of a Historical Man, Don Manuel Guzman Willis.*" *Cebu*, Vol. 5, No. 4., July and August, 1979. Indianapolis, Piso, Mexico.

Horn, Paul. "Black Hybrids on the Greenbriar." *Cattleman Magazine*, Fort Worth, Texas, Feb., 1973.
Horn, Paul. "A Century with Brahmans." *The Cattleman.*, Feb., 1978.
Horn, Paul. "Fifth Generation Cowman, Richard Forgason is President of ABBA." *The Cattleman*, Feb., 1978.
Horn, Paul. "The Reds from Brazil." *The Cattleman*, 1978.
Hudgins, Edgar. "A History of Brahman Cattle." *Brangus Journal*, May, 1977.
Hudgins, Edgar H. *Guzerat Brahman Cattle and Facts on Cross-Breeding.* Hungerford, Texas, 1934.
"Hudgins, Mr. Brahman, Fue la atracion en Cd. Valles." Article concerns recent Brahman show held in Cuidad Valles, Vera Cruz, Mexico. Property of Edgar Hudgins, Hungerford, Texas. June, 1980.
Hudgins, Edgar. Brahman breeder and ABBA Director. Letter to Joe A. Akerman regarding founding of Brangus breed. Hungerford, Texas, July 8, 1980.
Hudgins, Walter T. *Brahman Cattle.* Hungerford, Texas.
Hruska, Roman L. *Cattle Germ Plasm Evaluation Program.* Progress Report No. 7, Meat Animal Research Center, Clay Center, Nebraska, 1972.
Hybrid Vigor in Brahman-Hereford Crosses. Technical Monograph 1. Texas A&M Univ. College Station, Texas. May, 1964.
"Iberia Livestock Farm Studies on Brahmans, 1932-1954." Jeanerette, Louisiana. ABBA Papers, Houston, Texas.
"The Indian Cattle Which Have Had the Greatest Influence in the Development of the American Brahman." *News* (From P. R. Dept. Giezendanner Co., Houston, Texas based on information from Farmers Bulletin No. 1361 by Virgil Parr, Cot., 1923.) Property of ABBA, Houston, Texas.
"An Industry First, Hereford-Brahman Field Day Deemed Highly Successful." *Texas Hereford*, June, 1980.
"J. D. Hudgins Ranch." *Brahman Feeder*, XV. Houston, Texas, July, 1949.
"Jacobs Bull to Honduras." *Coastal Cattleman.* Beaumont, Texas. Aug. 1939.
Jacobs, William States. Description of a number of Brahman bulls in 1924, 1925 Importations. Property of ABBA, Houston, Texas. 1949.
Jacobs, William States. "History of Brahman Importations," *American Brahman Journal.* Houston, Texas. Feb., 1949.
Jacobs, William States. Letter to Gail Whitcomb, ABBA President, regarding Brahmans owned by Dr. Jacobs. Houston, Texas, Aug. 10, 1948.
Jacobs, William States. *The Romance of the Brahman.* ABBA Papers. Houston.
"Joe D. Hughes Ranch." *Brahman Breeder-Feeder.* Houston, Texas. June, 1944.

John, Roger McKinley. *A History of the Brahman Cattle Industry in Texas. Texas A&M thesis for Masters degree requirement (Property of ABBA, Houston, Texas.)*

Joshi, P.N. Superintendent, Darbar Dairy of Bhavnagar, India. Letter written to Walter J. Hudgins, April 9, 1938.

Joshi, N.R. and Phillips, Ralph W. An FAO Study. *Zebu Cattle of India and Pakistan.* Rome, March, 1953.

Joshi, P.N. Superintendent, Darbar Dairy Farm, Bhavnagar, India. Letter written to Walter Hudgins Sept. 8, 1939, regarding American Brahmans.

"Judging Brahman Cattle." A color slide set. Nasco. Fort Atkinson, Wisconsin, Property of ABBA.

Kelley, R.B. Progress Reports Nos. 2, 3 and 4 of the Council for the Scientific and Industrial Research, Commonwealth of Australia, *Zebu (Brahman) Cross Cattle and Their Possibilities in North Australia.* Sydney, 1936, 1938, 1940.

Kincaid, C.M. *Breed Crosses with Beef Cattle in the South.* USDA Southern Cooperative Bulletin No. 81, March, 1961.

Kirksey, George. "Report on Foot and Mouth Disease in Mexico." *American Brahman Journal,* Houston, Texas. Oct. 1948 and Jan., 1949.

Koch, Charles R. "Fiftieth Anniversary for a Centuries Old Breed." *The Brahman Journal.* Chaparral. Sealy, Texas, Feb., 1974.

Koger, Marvin. Professor of Animal Breeding and Genetics."Characteristics of Types and Breeds of Cattle in the Tropics." University of Fla., Gainesville, Fla., 1978.

Koger, Marvin; Cunha, Tony J. and Warnick, Alvin C. (editors).*Crossbreeding Beef Cattle, Series 2.* University of Fla., Gainesville, 1973.

Koger, Marvin. Livestock Geneticist and Professor of Animal Science, University of Fla. "Crossbreeding Zebu Cattle," article to appear in *Journal of Animal Science,* Sept., 1980.

Koger, Marvin; Peacock, Fentress; Kirk. W.G. and Crockett, J.R. "Heterosis Effects on Weaning Performance of Brahman-Shorthorn Calves." University of Fla. *Journal of Animal Science,* Vol. 40, No. 5, 1975.

Koontz, Henry Clay. "Early Red Brahman History," *Red Brahman Directory,* Placedo, Texas, 1976.

Koontz, Henry Clay. "HK Ranch History," *Red Brahman Directory,* 1976.

Koontz, Henry Clay and Dial, Emily. *History of Keeran Baker Herd.* Placedo, Texas, Oct. 23, 1968.

Ladd, Kevin. "The Story of the Fabulous Shanghai Pierce." *Texas Illustrated Magazine.*

Lambert, Teres. "Simbrah Brings Out Best of Two Breeds." *Simbrah Special.* American Simmental Assoc., Lindsborg, Kansas, July, 1980.

Lawson, J. E. Animal Geneticist. "Brahman X Hereford Hybrids Excel in Southern Alberta." *The Cattleman.,* 1970.

Lilley, Roy W. "Origin and History of Brangus." *Brangus Journal*, May, 1977.

Locke, Leon. President of ABBA. "American Brahman Cattle and Their Influence on World Beef Production." Address made to ABBA, 1966, Houston, Texas.

Locke, Leon. President of ABBA. "The Need for Improvement." Louisiana Association Field Day, June 8, 1967.

L.S.U. Research Studies. ABBA Papers, Houston, Texas.

Love, Clara. "History of the Cattle Industry in the Southwest." *The Southwestern Historical Quarterly*. The Texas State Historical Assoc., Austin, Texas, July, 1916.

McCormick and B.L. Southwell. "A Comparison of Brahman Crossbred Cattle" Tifton: Georgia Coastal Plain Experiment Station, 1957, *Journal of Animal Science*, Vol. 16.

McCully, Jack T. "Fat Stock Show and Livestock Exposition To Be One of Nation's Largest." *Coastal Cattleman*. Beaumont, Tx. Feb., 1938.

"The Ranch That Is Making History." McKellar Ranch brochure. Mount Pleasant, Tx. 1980.

"Marceaux, Aubrey J." *Brahman Breeder-Feeder. Dec. 1941.*

Martin, Margaret. "Florida," Brahman Breeder-Feeder. Houston, Aug. 1946.

Mendenhall, Mick (Brahman F_1 breeder). Letter concerning performance of Brahman-Hereford crosses near Rocky Ford, Colorado. Property of ABBA. Houston, Tx. Nov. 1970.

"Milby Butler Ranch," *Brahman Breeder-Feeder*. Houston, July, 1945.

Moore, R. L., Essig, H.N. and Smithson, L.J. "Influence on Breeds of Beef Cattle on Ration Utilization," *Journal of Animal Science*, 1975.

Muir, Forest Andrew. Professor at Rice University. Letter to G. Whitcomb, ABBA President, regarding transfer of Brahman cattle in 1860's. Houston, Dec. 11. 1967.

"M.S. Holsworth Ranch, A History," *Red Brahman Directory*, Property of ABBA. Houston.

Nighbert, E. M. (D.V.M.). "The Importance of Tick Eradication." *The Cattleman*. Fort Worth. Oct. 1914.

Nicholson, Arnold. "Texas' Own Sacred Cows." *The Cattleman*. Feb. 1943.

Oliver, Col. Arthur. British animal husbandry expert, Imperial Council of Agricultural Research. *A Brief Survey of the Imported Breeds of Cattle in India*. Bulletin No. 17. Govn't. of India Press. New Delhi: 1938.

"One Hundred Years in the Brahman Business." *Stockman/Farmer.*
Orts, Frank. Texas A&M University. "Carcass Characteristics of Brahman Crosses." *The Cattleman.* Fort Worth: Feb. 1971.
"Panamanians Buy Cattle." News Release Fla. Dept. of Agriculture and Consumer Services, Tallahassee: Sept. 1980.
Park, Jim. "Henry O. Partin, Pioneer Cattleman." *American Brahman Journal.* Houston: Sept. 1950.
Parr, Virgil V., animal husbandman. Bureau of Animal Industry, USDA. "Brahman (Zebu) Cattle." *Farmer's Bulletin* No. 1361. Washington, D.C. 1924.
Partin, Henry O. ABBA President and Fla. Brahman breeder, "Brahmans Playing Important Role in Increasing Beef Production." *The Cattleman.* 1950.
Patterson, R.E. Texas A&M. *Hybrid Vigor in Brahman-Hereford Crosses.* May, 1964.
Patton, Bruce. "Milestone for 3X." *The Victoria Advocate.* Victoria, Tx. Aug. 4, 1978.
Phillips, Ralph W. "World Distribution of the Major Types of Cattle." *Journal of Heredity* (Reprint) Washington, D.C. Sept.-Oct. 1961.
"Potter, (Mrs.) Palmer Brings Brahmans to Florida." *Florida Trend.* Tampa: Jan., 1976.
Preece, Hal. "The Cattle That Changed the West." *Real West.* May, 1968.
Preece, Hal. "Record Keeper, Margaret Watkins." *The Cattleman.* Fort Worth: Feb. 1970.
Randel, R. D. *Interrelationship of Endocrine and Physiological Events During the Estrous Cycle in Brahman Cattle.* Texas A&M University. Overton, Texas. 1980.
"Red Brahmans at Cherokee Ranch." *The Zebu Journal.* Cotulla, Tx. Sept. 1950.
"Red Brahman Bull Sells For Record Price." *The Brahman Journal.* Chaparral Publishing Co. Sealy, Tx. Feb. 1974.
"Red Brahmans on H. Cross Ranch." *The Zebu Journal.* Cotulla, Tx. Sept. 1950.
"Red Brahmans That Breed Red." *The Zebu Journal.* Cotulla, Tx. Sept. 1950.
Rhoad, A. O. "The Iberian Heat Tolerance Test," *Tropical Agriculture,* Vol. 21, 1944.
Rhoad, A.O. "New Blood and the American Brahmans." *Brahman Breeder-Feeder,* Houston: Sept. 1946.
Richardson, T.C. "Browsing in Brahmanland, *Farm and Ranch.* Aug., 1934.
Riggs, John K. "Brahman Cattle." *Phil Farmer.* Phillips Petroleum Co. Bartlesville, Okla. June 1948.

Rodeffer, Charles C. "Brahma Blood Brought Here by Cattlemen." Norfolk, Va. June 12, 1949.
Rouse, John E. *The Criollo, Spanish Cattle in the Americas.* University of Oklahoma Press. Norman: 1977.
Rouse, John E. *World Cattle.* Vols. I and II. University of Oklahoma Press. Norman: 1970.
Russell, Joan. Secretary of Public Relations, ABBA. "How the Cattle of India Came to America." *The Cattleman.* Fort Worth: May, 1958, and "Brahman Breed Helped by Teddy Roosevelt Intervention." *International Brahman Review*, April, 1960.
Russell, Joan. Secretary of Public Relations, ABBA "Brahmans in 35 Years Have World-Wide Influence." *The Florida Cattleman.* Kissimmee, Fla. Feb. 1959.
Russell, J.W. "Interview with James W. Sartwelle." Property of ABBA. Houston: April 14, 1959.
Sanders, Jim. Professor of Animal Science. Texas A&M University. "A History of the Zebu in Brazil." College Station, Tx. 1980.
"Santa Gertrudis Cattle." *Coastal Cattleman.* Beaumont: March, 1940.
Sartwelle, Caroline Olivia. Letter to P. T. Sartwelle of Comanche, Tx. regarding purchase of Brahman cattle. Dec. 2, 1920.
Sartwelle, J. D. Brahman breeder and ABBA Director. Letter regarding Sartwelle and Dinsmore family roles in early Texas cattle industry. July 29, 1980.
Sartwelle, J.W. Brahman breeder and ABBA president, "Brief History of Brahman Cattle." *The Cattleman.* Beaumont: July, 1949.
Sartwelle, J.W. "Brahmans Have Interesting, Unusual Background and Are World's Oldest Breed of Cattle." *Coastal Cattleman.* Beaumont: Dec. 1939.
Sartwelle, J.W. *Interesting Points About Brahman Cattle* (transcript of radio address made over KPRC, Houston, Oct. 14, 1939).
Sartwelle, J.W. Letter to D.M.K. Reddy of Cornell University concerning *Bos indicus* cattle. Nov. 17, 1924. Property of ABBA, Houston.
Sartwelle, J.W. "Officials of ABBA Make Official Trip to Cuba." *Coastal Cattleman.* Beaumont: Oct. 1937.
"Sartwelle's Own Pioneer Herd." *The Houston Post.* Feb. 3, 1946.
Schreiner, Charles III, "The Background and Development of Brahman Cattle in Texas," *The S.W. Historical Quarterly*, LII, April, 1949.
Schronk, Wendell, ABBA Exec. Vice-president. *ABBA Partically Refunds $186,000 Research Project*, Property of ABBA, Houston.
Schronk, Wendell. Address made at Brahman/Angus F_1Field Day. Brooksville, Fla. July 19, 1980.
Schronk, Wendell. Address made at Hereford/Brahman F_1 Field Day held at Comanche, Texas. May 10, 1980.

Schronk, Wendell. "Bill Coman - A Man as Large as the Country He Ranches." *The Brahman Journal.* Sagebrush Publishing Co., Inc., Sealy, Tx. May, 1976.
Schronk, Wendell. "Garcia and 1946 Importation," Interview with Esteban Garcia, Sr., Property of ABBA, Houston.
Schronk, Wendell. "Know the 'Flip Side' of Your Simbrah: The Brahman." *Simmental Shield.* Bozeman, Mont. July, 1980.
Schronk, Wendell and Ed Cullers, ABBA Director. "Marketing Trip to Zambia and Zimbabwe," ABBA Release. Houston: Aug. 5-7, 1980.
Schronk, Wendell. "Mongrels from the Melbourne Zoo." *The Brahman Journal.* Sagebrush Publishing Co., Inc. Sealy, Texas: Feb. 1980.
"Shanghai Pierce - Colorful Texas Longhorn King." *The Houston Chronicle. Aug. 30, 1931.*
"Shares Brahman Monolopy with Friends." Brahman Breeder-Feeder. Houston: 1948.
Simons, Henry M. (Ed.) *Proceedings of the American Brahman Centennial. Charleston, S.C. 1949.*
Simpson, James R. Extension Marketing Economist, Univ. of Fla. "Exporting Florida Beef Breeding Cattle; Potential, Prospects and Problems." Florida Cattleman, FCA. Kissimmee, Fla: Aug. 1980.
Sims, Grover J. "U.S. Cattle To Improve South American Herds." Foreign Agriculture. Vol. XIX, No. 3. Foreign Agricultural Service of USDA. Washington, D.C. March, 1955.
Smith, Carlos. "Zebu." *Brahman Breeder-Feeder.* Houston: June, 1949.
"Sonora Ranch Studies." Texas A&M College, 1920-1929. ABBA Papers. Houston, Texas.
"Southwest Bull Classic, 1980." *The Brahman Journal.* Feb. 1980.
"Spirit of '46 Sale." *The Brahman Journal.* Sealy, Tx. Jan. 1980.
"Stanley Kubela Prefers Red Brahmans." *The Zebu Journal.* Sept. 1950.
"Steer Our Way." Ad in *The Cattleman* sponsored by ABBA on Brahman hybrids showing cut-out of 1978 steer. Fort Worth: July, 1980.
Stone, J.W. "J. M. Frost Blazes the Trail." Article owned by Vernon W. Frost published earlier in *Brahman Breeder-Feeder.* Houston.
Taylor, J.M. Letter written Sept. 25, 1875 and sent to A. R. Hudgins of Chappell Hill, Texas regarding birth of Brahman calf. Photostat of letter owned by ABBA, Houston.
"Texas Foot and Mouth Disease Outbreak." *The Cattleman*, Oct. 1924.
"Thirty-Third Annual Brahamn Section." *The Florida Cattleman. FCA. Kissimmee, Fla. March, 1976.*
Thompson, Fred. "Beefmasters Meet Industry's Critical Needs." The Florida Cattleman. FCA. Kissimmee, Fla.: Jan. 1976.
"3X Ranch Family Tree." *Brahman Breeder-Feeder.* Houston: Jan. 1947.
"Three Generations of Brahman Breeders, Rainbow 'S' Ranch." *Red Brahman Directory.* Property of ABBA. Houston.

Transfer Records Into Countries Outside U.S.A. (1969-1980) ABBA, Houston.
"Trio of Brazilians Here for Fat Stock Show." *Houston Post.* 1946.
Turner, J.W. "Genetic and Biological Aspects of Zebu Adaptability." L.S.U. Baton Rouge, Louisiana. (Presented at symposium at meeting of ASAS, Feb. 5, 1979 in New Orleans).
"United States Exports of Breeding Cattle, 1945-1954; 1965-1977." USDA U.S. Printing Office. Washington, D.C.
Victoria (The) *Sequiscentennial Inc.* Victoria, Texas, 1974.
"W.H. Stuart Salutes ABBA's 50th." *The Brahman Journal.* Chaparral Press. Sealy, Tx. Feb., 1974.
Walter, John. "What Is a Simbrah?" *Simmental Shield.* Bozeman: American Simmental Assoc., July, 1980.
Whitcomb, Gail, President ABBA. "Observations on Brahman Cattle." *American Brahman Journal.* Houston, Feb. 1949.
Weil, Black, Blutstein, Martindale, McMorris, Merry, Munson and Townsend.
Area Handbook for Venezuela. The American Univ., Washinton, D.C. 1971.
Wideman, Don. "First American Beef Breed." *Stockman/Farmer.*
Willis, Manuel Guzman. *The History of the Brahman Race in Mexico.* Tampico, Mexico: 1945.
Y'turria, Frank D. *Bos indicus, A Brief History and Description of the Different Breeds* (Second Edition) Property of ABBA. 1973.
Zipf, Bill. "Brahmans Were Most Popular Livestock Display at Fair." *The Columbus Dispatch.* Columbus, Ohio: Sept. 2, 1951.

INTERVIEWS

Allen, Steve. Manager. Santa Elena Ranch. Madisonville, Tx. Jan. 10, 1981.
Banks, Monty. General Manager, McKellar Ranch, Mt. Pleasant, Texas. August 1, 1980.
Beville, Herman, Brahman breeder and former ABBA Director. Interview Bushnell, Fla., July 30, 1980.
Brockett, Joe Ed. Editor, The *Brahman Journal.* Interview. Houston, Texas, August 2, 1980.
Burns, W. C. "Cal". Director, USDA Beef Cattle Research Cetner. Brooksville, Fla., July 19, 1980.
Carlton, Greg. Animal scientist, United States Sugar Corp. Interview, Clewiston, Fla., June 2, 1980.

Davis, Avon. Brahman breeder. Interview. A. D. Ranch, Chowchilla, Calif. Aug. 5, 1980.

Dial, Emily. Brahman breeder. 3X Ranch, Placedo, Texas, Aug. 6, 1980.

Dial, Joe. Brahman breeder and ABBA Director. 3X Ranch, Placedo, Texas. 9/29/79; 2/14/80; 8/6/80; 6/20/81.

Dobbins, Claude. Livestock marketing specialist, USDA. (Retired). Advisor to C. Clausen Steampship Co., Alexandria, Va., July 30, 1980.

Frost, Vernon. Brahman breeder. Pecan Acres Ranch, Houston, Texas. Aug. 2, 1980.

Gamble, Marvin. Brahman breeder, Perry, Florida, May 10, 1979.

Gayden, Harry P. Former Executive Secretary of ABBA. Director of ABBA. Houston, Texas. Feb. 22, 1980.

Hammond, M. E. Former President of ABBA, Manager of W. H. Stuart Ranch, Inc. Brooksville, Fla. July 19, 1980.

Hudgins, Edgar. Brahman breeder. Director of ABBA. Hungerford, Texas. May 29 and 30, 1980; July 8, 1980; July 9, 1981.

Johnson, Bryon. Former Purebred Herd Mgr. for Norris Cattle Co., Ocala, Jan. 14, 1981.

Kinner, William and Leah. Romano Red breeders. Interview. Madison, Fla. July 21, 1980.

Koger, Marvin. Professor of Animal Science at University of Florida. Gainesville, Fla. Interview. July 2, 1980.

Koontz, Henry Clay. Brahman breeder and owner of HK Ranch, Placedo, Texas. Aug. 1, 1980.

McCormick, W.C., PhD. Coastal Plain Experiment Station. Tifton, Georgia: Aug. 20, 1981.

Marceaux, Duffy. Brahman breeder. Kaplan, Louisiana. July 30, 1980.

Martin, Roy. Secretary of PAZA and Brahman breeder. Cotulla, Texas. Interview. May 27, 1980.

Pace, James. Former Professor of Animal Science. University of Florida, Gainesville, Fla. July 29, 1980.

Partin, O. L. Brahman breeder and ABBA Director. Kissimmee, Fla. April 10, 1976.

Patterson, Lewis. Brahman breeder and ABBA Director. Concord, N.C. July 25, 1981.

Peeples, Billy. Cattle breeder from Moore Haven, Fla. Brooksville, Fla. July 19, 1980.

Pingrey, Dave. President of Amerian Angus Association. Angus-Brahman Field Day. Brooksville, Fla. July 19, 1980.

Pritchard, Lee. Director of ABBA Services. American Brahman Breeders Assoc. Houston, Texas. June 10, 1981.

Ramsey, Rick. Completing research on Frank Buffaloe. Houston, Texas. May 29, 1980.

Rogillio (Forbes), Margie. Brahman breeder. Interview. Greenwood Springs, Louisiana, July 31, 1980.
Rouse, John E. Cattleman and author. Interview. Saratoga, Wyoming. Aug. 15, 1980.
Sanders, James. Professor of Animal Science, Texas A&M Univ. Interview. College Station, Texas. June 30, 1980.
Sartwelle, James. Director of ABBA. Brahman breeder. Interview. Houston, Texas. Feb. 22, 1980.
Schronk, Wendell. Executive Secretary of ABBA. Interview. Houston, Texas. Feb. 22, 1980.
Sharman, E. C. "Buck". USDA. Interview. Hyattsville, Md. Feb. 8, 1980.
Smoak, Ronnie. Hereford breeder and professor of agriculture at North Florida Junior College, Madison, Fla. June 20, 1980.
Stack, G. T. "Gerry". Brahman breeder and exporter. Interview. Tampa, Fla. Aug. 4, 1980; Feb. 10, 1981; May 5, 1981; Oct. 28, 1981.
Stiles, John. Director of Marketing. Fla. Dept. of Agriculture and Consumer Services. Interview. Tallahassee, Fla. Aug. 14, 1981.
Tucker, Andy. Pres. Eastern Brahman Assoc. Interview. Brooksville, Fla. July 19, 1980.
Tucker, Gilbert. Director of ABBA and Brahman breeder. G. A. Tucker and Sons. Interview. Brooksville, Fla. July 19, 1980.
Turner, William. Professor of Animal Science, Louisiana State Univ. Interview. Baton Rouge, Louisiana. June 27. 1980.
Watkins, Margaret. Former Recording Secretary, ABBA. Brahman breeder. Interview. Edna, Texas. June 30, 1980.
White, J. T. Brahman breeder and former ABBA President. Herne, Texas. Nov. 12, 1980.
Wright, Betty. McKellar Ranch employee. Interview. Mt. Pleasant, Texas. Aug. 2, 1980.

Footnotes

Chapter I

1. Lloyd, Clyburn, Former Director of Education and Information, ABBA, Unpublished manuscript on the American Brahman, Houston: 1955, pp. 42-55; J. W. Sartwelle, Former ABBA President and Director (Letter written to D.M.K. Reddy, Cornell University, November 17, 1979 containing definition of Brahman); Edgar Hudgins, ABBA Director and Former ABBA President, interview, Hungerford, Tx., May 29, 30, 1980.

2. *Ibid.*, N. R. Joshi and Ralph W. Phillips, FAO Agricultural Studies, *Zebu Cattle of India and Pakistan, Rome, Italy: March, 1953, pp. 14-19; Col. Sir Arthur Oliver, Animal Husbandry Expert, Imperial Council of Agricultural Research, A Brief Survey of Some of the Important Breeds of Cattle in India,* Government of India Press, New Delhi: 1938, pp. 1-8.

3. Wendell Schronk, ABBA Executive Vice-president, interview, Houston, February 23, 1980.

4. Edgar Hudgins, "A History of Brahman Cattle," *Brangus Journal*, May, 1977, p. 48; Edgar Hudgins, *op. cit.* interview, May 29, 1980. Mrs. S. C. Border, "Brahmans . . . Bred Through the Centuries for the South of Today," *Coastal Cattleman*, Aug., 1940, p. 18.

5. James W. Sartwelle, *op. cit.* (information included in letter to D.M.K. Reddy), ABBA Papers, Houston: p. 1; N. R. Joshi and Ralph W. Phillips, *op. cit., pp. 23-39; Col. Sir Arthur Oliver, op. cit.* p. 20; Virgil V. Parr, *Brahman (Zebu) Cattle*, USDA, Farmers' Bulletin No. 1361, Washington, D.C. 1923, pp. 8-9; James O. Sanders, Texas A&M, "History and Development of Zebu Cattle in the United States," *Journal of Animal Science*, Vol. 50, No. 6, 1980, pp. 1190-1191 (quoted from Col. Sir Arthur Oliver, *op. cit*, p. 20).

6. Clyburn, *op. cit.*, p. 34.

7. *Enclyclopedia Britannica*, I Volume, 1976. Edgar Hudgins, *Guzerat Brahman Cattle and Facts on Cross-Breeding*, (pamphlet published by Hudgins), pp. 1-2. Hungerford, Tx., 1934; Joshi and Phillips, *op. cit.*, pp. 24-25.

8. Joshi and Phillips, *op. cit.*, p. 33.

10. *Ibid.*, pp. 153-157; *Encyclopedia Britannica*, Vol. I, 1976.

11. *Ibid.*, p. 157; Frank D. Y'turria, *Bos Indicus, a Brief History and Description of Some of the Different Breeds*, 1973 (ABBA Papers, Houston).

12. *Ibid., p. 159; Clyburn, op. cit.*, pp. 39-40; Frank D. Y'turria, *op. cit.*

13. Frank D. Y'turria, *op. cit.*; Joshi and Phillips, *op. cit.*, pp. 163-166.

14. Joshi and Phillips, *op. cit.*, p. 166.

15. Joe Dial, ABBA Director, interview, 3X Ranch, Placedo, Texas, Feb. 14, 1980.

16. Joe Dial, transcript of speech given on history of Brahman, ABBA Papers, Houston: 1980.

17. Joshi and Phillips, *op. cit.*, pp. 118-120; Clyburn, p. 36.

18. Clyburn, pp. 36-37.

19. *Ibid.*, p. 47.

20. Joshi and Phillips, *op. cit.*, p. 126.

21. "Texas Nelore Importation Largest in Last Forty Years," *The Brahman Journal,* Eddy, Texas, August, p. 20.

22. Joshi and Phillips, *op. cit.*, pp. 97-102.

23. Sanders, *op. cit.* p. 1196. J.D. Sartwelle, ABBA Director, interview, Dallas: Oct. 18, 1980.

24. Sanders, *op. cit.*, p. 1193.

25. Frank D. Y'turria, *op. cit.*

26. Dial, Transcript of Speech, *op. cit.*

27. E. C. "Buck" Sharman, USDA, interview. Hyattsville, Md., Feb. 8, 1980.

28. Dial, *op. cit.*; J. Frank Dobie, *The Long Horns,* Little Brown and Company, Boston: 1941, pp. 26-42 and 43-69. J. L. Baughman, "Climate, Cattle and Crossbreeding Beef and Milk Production in the Tropics and Subtropics," *The Texas Journal of Science,* June 30, 1951, pp. 257-258.

29. Baughman, *op. cit.*, p. 258.

30. *Ibid.*, p. 259.

31. Dobie, *op. cit.*

32. Harry Gayden, Former Executive Secretary of ABBA, interview, Houston, Tx., Feb. 22, 1980; Edgar Hudgins, *op. cit.*; Wendell Schronk, *op. cit.*

33. Joe Akerman, *Florida Cowman, A History of Florida Cattle Raising.* FCA, Kissimmee, 1976, pp. 232-234. Letter from J. M. Taylor of Quinan, Tx., to A. R. Hudgins of Chappel Hill, Tx. Sept. 25, 1976; Letter in ABBA Papers regarding transfer of *Brahma bull, 1869, Houston; Chris Emmett, Shanghai Pierce, A Fair Likeness,* University of Oklahoma Press, Norman: 1953, p. 304.

34. J. W. Sartwelle, letter to D.M.K. Reddy, *op. cit.*, p. 1; Clyburn, *op. cit.*, p. 57.

35. *Ibid.*; J. D. Sartwelle, *op. cit.*

36. "ABBA - Twenty-five Years of Progress," *Brahman Breeder-Feeder,* XV, Jan., 1949, pp. 9-10. "ABBA Minutes, 1924," ABBA Headquarters, Houston.

37. - 38. "Standard of Excellence for American Brahman Cattle," *The American Brahman,* ABBA, Houston, 1980.

Chapter II

1. Lloyd, Clyburn, Former ABBA Director Information and Education, *Development of the American Brahman* (unpublished manuscript), April 25, 1955, pp. 27-28; John K. Riggs, "Brahman Cattle," *Zebu Journal,* June, 1948, p. 5.

2. *Ibid.*, p. 28; P. N. Joshi, Superintendent Darbar Herd of Bhavnagar, India. (letter to Walter Hudgins, ABBA Director), Hungerford, Texas, Feb. 1938.

3. P. N. Joshi and Ralph W. Phillips, cited by Clyburn in *Development of the American Brahman, op. cit.*; John Francis, "Definition and Use of Zebu, Brahman or *Bos indicus* Cattle" *Nature,* Vol. 207, Sydney, Australia, July 3, 1965, p. 2.

4. Clyburn, *op. cit.*, p. 29.

5. *Ibid.*, pp. 40-41.

6. *Ibid.*, p. 40. Ralph W. Phillips. "The Cattle of India", *The Journal of Heredity,* Washington, D.C. Vol. XXXV, September, 1944.

7. M. K. Reddy, Cornell University, cited by Lloyd Clyburn in *Development of the American Brahman, op. cit.*, pp. 29-30.

8. Clyburn, *op. cit.*, p. 29. P.N. Joshi, *op. cit.*, pp. 1 and 2.

9. Clyburn, *op. cit.*, p. 41.

10. "Trio of Brazilians Here for Fat Stock Show," *Houston Post.* July, 1946.

11. Roger McKinley John, *A History of the Brahman Cattle Industry in Texas* (a Thesis, Texas A&M), College Station: 1966, p. 18.

12. "Trio of Brazilians Here for Fat Stock Show," *op. cit.*

13. John, *op. cit.*; J. L. Baughman, "Climate, Cattle and Crossbreeding Beef and Milk Production in the Tropics and Subtropics." *The Texas Journal of Science,* No. 2, June, 1951, p. 261.

14. Charles Schreiner, III, "The Background and Development of the Brahman Cattle in Texas," *The Southwestern Historical Quarterly, LII,* April, 1949, p. 431.

15. John, *op. cit.*, pp. 18-19.

16. *Ibid.*, p. 19. J. W. Sartwelle, "Brahmans Have Interesting, Unusual Background. . ." *Coastal Cattleman,* C.C.A., Beaumont: Dec. 1939, p. 6; Clyburn, *op. cit.*, pp. 19-20.

17. Andrew Forest Muir; Professor Rice University (letter to ABBA President Gail Whitcomb regarding early Brahmans), Dec. 11, 1967.

18. John, *op. cit.*, p. 20; William Jacobs "History of Brahman Importations," *American Brahman Journal,* Feb. 1949.

19. Clyburn, *op. cit.*, pp. 42-45; Jacobs, *op. cit.*; John, *op. cit.*, p. 22; Hal Preece, "The Cattle That Changed the West," *Real West,* May, 1968.

20. Charles Koch, "Fiftieth Anniversary . . ." *Brahman Journal,* Feb. 1974, p. 38; Clyburn, *op. cit.*; *The Victoria Sequiscentennial, Inc.*, Victoria, Texas, 1974; Joe Dial

Speech, "History of the American Brahman," Houston: 1979.

21. J. L. Baughman, op. cit., p. 262; ABBA Reference Materials, ABBA office, Houston, Texas.

22."A. B. Lawrence Ranch," *The Houston Post*, Feb. 3, 1946; "Brahmans in Louisiana -The Story of Brahman Cattle at L.S.U. and in the Baton Rouge Area," *Gulf Coast Cattleman*, Louisiana Cattlemen's Assoc., Nov. 1948.

23. Clyburn, *op. cit.*, p. 56.

24. Joe Akerman, "Rustling, Dipping, Fencing, and the End of an Era," *Florida Cowman*, FCA, Kissimmee: 1976, p. 232.

25. Akerman, *op. cit.*, p. 250; *Sunland Tribune*, Tampa: March 1, 1879.

27. John, op. cit., pp. 20-21; Joe Dial Speech, *op. cit.*; Paul Horn, "A Century With Brahmans," *The Cattleman*, *op. cit.*, p. 50; Henry Clay Koontz and Emily Dial, *History of Keeran-Baker Herd*. ABBA materials, Oct. 23, 1968. "A Brief History of Brahman Cattle in the United States," *American Brahman Journal*, April, 1946, p. 3.

28. Clyburn, *op. cit.*, pp. 44-45; J. L. Baughman, *op. cit.*, p. 262. John, *op. cit.*, pp. 23-24.

29. John, pp. 24-25; C. M. Frost, "The First Brahman Herd in Texas," *The Zebu Journal*, August, 1949, pp. 9-12; Vernon Frost, Brahman breeder, Pecan Acres Ranch (interview), Houston: Aug. 2, 1980; James O. Sanders, *op. cit.*, p. 1195.

30. Vernon Frost, *op. cit.*

31. *Ibid.*

32. Preece, *op. cit.*, p. 3; Joan Russell, Sec. Public Relations, ABBA, "Brahman, Breed Helped by Teddy Roosevelt Intervention," *The International Brahman Review*, April, 1960, p. 12; Edgar Hudgins, "A History of Brahman Cattle," *Brangus Journal*, May, 1977, p. 15.

33. John, *op. cit.*, p. 25.

34. Clyburn, *op. cit.*, p. 45; Joan Russell, *op. cit.*, p. 12.

35. Clyburn, *op. cit.*

36. J. W. Sartwelle, "Brief History of Brahman Cattle," *The Cattleman*, July, 1939; "Milby Butler Ranch," *Brahman Breeder-Feeder*, Houston: July, 1945, pp. 42-43.

37. Joe Brockett, "Family's Brahman Story," *The Brahman Journal*, August, 1978, pp. 5-6.

38. James Sanders, *op. cit.*, p. 1195; Charles Koch, *op. cit.*, p. 39.

39. John, *op. cit.*, p. 26.

40. *Ibid.*; Clyburn, *op. cit.*, p. 46.

41. Sanders, *op. cit.*, p. 1195; J. L. Baughman, *op. cit.*, p. 262. J. W. Sartwelle, "Brief History of Brahman Cattle," *op. cit.*, July, 1939; John, *op. cit.*, p. 26.

42. John, *op. cit.*, p. 28.

43. Clara M. Love, "History of the Cattle Industry in the Southwest," *The Southwestern Historical Quarterly*, pp. 1-3; T. R. Havins, "Texas Fever," *The Southwestern Historical Quarterly*, Austin: July, 1948, pp. 147-162; Dr. E. M. Nighbert, "The Importance of Tick Eradication," *The Cattleman*, Fort Worth: Oct., 1914, pp. 3-4.

44. T. R. Havins, *op. cit.*, pp. 160-161.

45. Chris Emmett. *Shanghai Pierce: A Fair Likeness*, Norman: University of Okla. Press, 1953, pp. 303-304.

46. John, *op. cit.*, *p. 109; Edgar Hudgins, Brahman Breeder and ABBA Director, Hungerford, Tx. July 18, 1980; J. L. Baughman, op. cit.*, p. 263. T. R. Havins, *op. cit.*

47. *Emmett, Shanghai Pierce, A Fair Likeness* (cited from Bureau of Animal Industry, Department of Agriculture, 1909), p. 303.

48. *Ibid.*, p. 304.

49. Clyburn, *op. cit.*, p. 47; Hudgins, *op. cit.*

50. Koch, *op. cit.*, p. 39.

51. Emmett, *op. cit.*, pp. 301-302; John, *op. cit.*, p. 29 (cited from David Fairchild, *The World Was My Garden: Travels of a Plant Explorer*, New York: Charles Scribner and Sons, 1938, p. 225).

52. *Empty Tea Cup Fills Texas with Brahmas*. 1936-1937 (ABBA Materials, *op. cit.)* p. 7.

53. *Ibid.*, pp. 8-9; John, *op. cit.*, p. 30.

54. *Empty Tea Cup Fills Texas with Brahmas*, pp. 9-11; Clyburn, *op. cit.*, p. 47.

55. *Empty Tea Cup Fills Texas with Brahmas*, *op. cit.*, pp. 8-9.

56. *Ibid.*, pp. 9-10; Emmett, *Shanghai Pierce, A Fair Likeness*, *op. cit.*, p. 307.

57. Mayer L. Halff, (copy of letter written by New York attorney Mayer L. Halff addressed to Chris Emmett which appeared in *The Cattleman*, Nov., 1950); Clyburn, *op. cit.*, p. 48; Sam T. Cutbirth (letter to Dr. Burch H. Schneider, Sept. 13, 1949 concerning disposition of Borden-O'Connor Brahmans), ABBA Materials, *op. cit.*

58. Emmett, *Shanghai Pierce, A Fair Likeness*, p. 309.

59. Hudgins, "A History of Brahman Cattle," *op. cit.*, p. 15.

60. Clyburn, *op. cit.*, pp. 49-50.

61. Kirby Cunningham, "The Sartwelle Brothers, Texas Ranching Pioneers," *The Cattleman*, Feb., 1971, pp. 13-26; John, *op. cit.*,p. 74.

62. Horn, "A Century with Brahmans", *op. cit.*, p. 50; Catherine Barton, "Visiting the 3X Brahman Ranch", *Brahman Breeder-Feeder*, May, 1946, p. 27-34; "Mrs. Keeran Stocked Her Ranch with *Sacred Cows*", *Brahman Breeder-Feeder*, April, 1947, p. 31.

63. Cutbirth, *op. cit.*

64. Hudgins, *op. cit.*

65. Marvin Gamble, Brahman breeder (interview), Perry, Florida, May 10, 1979.

66. Clyburn, *op. cit.*, p. 49; A. P. Borden, "Report to American Breeders Society, 1908," (copy), ABBA Papers, Houston, Texas.

67. William States Jacobs, "History of Brahman Importations," *The American Brahman*, Oct., 1955, p. 10; Edgar Hudgins, ABBA Director and Brahman breeder, interview, Hungerford: July 8, 1980.

68. Jack T. McCully, "Fat Stock Show and Live Stock Exposition To Be One of Nation's Largest," *Coastal Cattleman*, Feb., 1938, p. 1.

69. Clyburn, *op. cit.*, p. 50.

70. *Ibid.*

71. "The Story of the American Brahman Breed," *American Brahman Cattle*, (publication released by ABBA, Houston, Texas), p. 3; Preece, *op. cit.*, p. 4; Clyburn, *op. cit.*, p. 50; Koch, *op. cit.*, pp. 39-40.

72. John, *op. cit.*, p. 41.

Chapter III

1. Roger McKinley John, Texas A&M, *A History of the Brahman Cattle Industry in Texas* (a Thesis, Texas A&M), College Station: 1965, p. 55.

2. *Ibid.*, Lloyd Clyburn, Director of Information and Education ABBA, Unpublished manuscript on American Brahman, Houston, 1955, p. 56; Kirby Cunningham, "The Sartwelle Brothers: Texas Ranching Pioneers." *The Cattleman*, Feb., 1971, p. 31.

3. Cunningham, *op. cit.*

4. Joan Russell, Interview with James W. Sartwelle, Palacios: April 14, 1958, p. 1.

5. *Ibid.*, pp. 2-3.

6. *Ibid.*, pp. 2-3.

7. *Ibid.*, p. 3.

8. *Ibid.*, p. 4.

9. *Ibid.*, p. 3.

10. *Ibid.*, pp. 3-4.

11. *Ibid.*, p. 4.

12. John, *op. cit.*, p. 56.

13. ABBA Minutes, 1924, ABBA office, Houston; Clyburn, *op. cit.*, p. 56.

14. John, *op. cit.*, p. 78.

15. ABBA Minutes, 1924, Houston.

16. Russell, *op. cit.*, p. 5; John, *op. cit.*, pp. 57-58.

17. William States Jacobs, "History of Brahman Importations," *American Brahman Journal*, Houston: Feb., 1949.

18. Edgar Hudgins, ABBA Director and former ABBA President (interview), Hungerford, Texas. June 6, 1980.

19. Russell, *op. cit.*, p. 4; James D. Sartwelle, Sr., ABBA Director and Brahman breeder (interview), Houston: July 8, 1980.

20. John, *op. cit.*, p. 58; Russell, *op. cit.*, p. 5.

21. Russell, *op. cit.*, p. 5.

22. *Ibid.*; ABBA Minutes, 1924, *op. cit.*

23. ABBA Minutes, 1924, *op. cit.*

24. Joan Russell, *op. cit.*, pp. 1-4.

25. Hudgins, *op. cit.*

26. ABBA Minutes, 1924, *op cit.*

27. Clyburn, *op. cit.*, p. 60; ABBA Minutes, 1924, *op. cit.*

28. Clyburn, *op. cit.*, p. 61; Jack T. McCully, "Fat Stock and Live Stock Exposition To Be One of Nation's Largest," *Coastal Cattleman*, Beaumont: Feb., 1938, pp. 5-6; Edgar Hudgins, *op. cit.*

29. Russell, *op. cit.*, p. 6.

30. ABBA Minutes, 1924, *op. cit.*

31. Russell, *op. cit.*

32. ABBA Minutes, 1941 (ABBA Transfer Records, 1924-1941), ABBA Office, Houston.

33. Hudgins, *op. cit.*; Clyburn, *op. cit.*, p. 61.

34. Clyburn, *op. cit.*, pp. 62-63.

35. ABBA Mintues, 1924-1948; *op. cit.*; Hudgins, *op. cit.*

36. Hudgins, *op. cit.*; ABBA Minutes, 1931-1947, *op. cit.*

37. ABBA Minutes, 1939, *op. cit.*

38. ABBA Minutes, 1939-1940, *op. cit.*

39. ABBA Minutes, 1941, *op. cit.*

40. Esteban Garcia (transcript of interview by Wendell Schronk, ABBA Exec. Secretary); Roy Martin, PAZA Secretary (interview by Joe A. Akerman), Cotulla, Tx., May 27, 1980; ABBA Mintues, 1946, *op. cit.*

41. ABBA Minutes, 1946-1980, *op. cit.*

42. Harold Preece, "Record Keeper . . . Margaret Watkins," *The Cattleman*, Feb. 1970.

43. *Ibid.*

44. ABBA Minutes, 1941-1949, *op. cit.*; ABBA Transfer Records, *op. cit.*

45. "Dr. William States Jacobs . . . Shares Brahman Monopoly with Friends," *American Brahman*, Houston: Sept. 1948, pp. 15-16.

46. ABBA Minutes, 1958, *op. cit.*

47. Gail Whitcomb, ABBA President, "Observations on Brahman Cattle," *American Brahman Journal*. Houston: Feb. 1949.

Chapter IV

1. Joe Dial, ABBA Director, (transcript of speech, "History of Brahman Cattle Registered in ABBA), 1980, ABBA Papers, Houston, p. 4.

2. Roger McKinley John, *A History of the Brahman Cattle Industry in Texas (a Thesis completed at Texas A&M: May, 1966 [Quote cited from interview with James W. Sartwelle, Houston: June 11, 1964), pp. 77-78.*

3. Ibid.

4. Edgar Hudgins, ABBA Director (interview), Hungerord, Texas, Feb. 4, 1980; John, *op. cit.*, pp. 77-78.

5. Lloyd Clyburn, Director of Information and Education, ABBA, Unpublished manuscript on American Brahman, Houston: 1955, p. 51.

6. Joe Ed Brockett, "Family's Brahman Story," *The Brahman Journal*, August, 1978, p. 39.

7. Hudgins, *op. cit.*

8. Dr. W. S. Jacobs, "A History of Brahmansfrom 1924 and 1925 Importations," part of ABBA Papers, Houston: 1949, p. 1.

9.*Ibid.*, p. 2.

10.-11. *Ibid.*, Hudgins, *op. cit.*

12. Jacobs, *op. cit.*, p. 3; Dial, *op. cit.*

13. James D. Sartwelle, Sr., ABBA Director (Letter addressed to Joe A. Akerman concerning lineage of cattle owned by Sartwelle Bros.), July 29, 1980.

14. Jacobs, *op. cit.*, p. 2.

15. *Ibid.*; Hudgins, *op. cit.*

16. Jacobs, *op. cit.*, pp. 5-6.

17. *Ibid.*, p. 1.

18. *Ibid.*, p. 3.

19.-20. *Ibid.*, p. 4.

21. Hudgins, *op. cit.* Sartwelle, *op. cit.*

22.-23. Brockett, *op. cit.*, p. 39; Hudgins, *op. cit.*

24. "Hudgins, Mr. Brahman, Fue la Atracion en Cd. Valles," *Cebu*, Valles, Mexico. 1980, pp. 5-15.; Edgar Hudgins, *op. cit.*

25. Hudgins, *op. cit.*; Dr. Richard L. Forgason, former ABBA President (Transcript of speech given at Brahman Field Day, July 23, 1976), ABBA Papers, Houston.

26. Forgason, *op. cit.*, p. 3.

27. Hudgins, *op. cit.*

28. Edgar Hudgins, Letter received by Joe A. Akerman, July 7, 1980.

29. Forgason, *op. cit.*, p. 5.

30.-32. *Ibid.*, pp. 5-6.

33. *Ibid.*, p. 6.

34.-35. *Ibid., p. 6-7; Hudgins, op. cit.*

36. Hudgins, *op. cit.*

37. Leon Locke, "Halter-Breaking and Gentling Calves Important Part of Brahman Operation," *International Brahman Review, XXIX*, Sept., 1963, p. 10.

38. Forgason, *op. cit.*, p. 9.

39. Joe Dial, ABBA Director, interview at 3X Ranch, Placedo, Texas, August 8, 1981.

40. Hudgins, *op. cit.*, Aug. 15, 1981.

41. *Ibid.*

42. Dial, Transcript of Speech, *op. cit.*, U. S. Sugar Corporation ad appearing in *The Brahman Journal*.

43. James D. Sartwelle, ABBA Director and General Partner Startwelle Brahman Ranch, Ltd., interview, Houston: July 30, 1980.

44. Sartwelle, *op. cit.*, Letter received by Joe A. Akerman, July 29, 1980; Jacobs, *op. cit.*

45. Duffy Marceaux, Brahman breeder, (interview), Kaplan, Louisiana, July 30, 1980.

46. Sartwelle, *op. cit.* (Information collected by James D. Sartwelle, Jr.).

47. Byron Johnson, former Norris Purebred Herd Manager (interview), Ocala: Jan. 14, 1981.

49. Hudgins, *op. cit.*; Sartwelle, *op. cit.*

50. Hudgins, *op. cit;* Jim Park., "Henry O. Partin Cattleman," *American Brahman*, Sept. 1950, pp. 11-12.

51. Joe A. Akerman, "Florida's First Cowmen," *The Florida Cowman*, FCA, Kissimmee, pp. 1-16; pp. 249-250.

52. Hudgins, *op. cit.*, Sartwelle, *op. cit.*

53. "Mr. Henry and the Boys," *The Tampa Tribune*, Sept. 22, 1957; "Mr. Henry's Empire," *Saturday Evening Post*, Jan. 7, 1956; "Hits Jackpot," *Tampa Sunday Tribune*, Aug. 24, 1952.

54. "*Emperor* Leased to Texan," *The Stockman*, March 1950, p. 2.

55. M. E. Hammond, Former ABBA President and Manager of W. H. Stuart Ranch, Inc. (interview), Bartow, Fla., July 19, 1980.

56. Jacobs, *op. cit.*

57. *Ibid.*, p. 7.
58. *Ibid.*, p. 1.
59. *Ibid.*, p. 2; Hudgins, *op. cit.*
60. Jacobs, *op. cit.*, pp. 9-10.
61. *Ibid.*, p. 10; Rick Ramsey (interview, relative of Frank Buffaloe completing research of Bufffaloe herd), Houston: May 29, 1980.
62. Jacobs, *op. cit.*, p. 10.
63. *Ibid.*, Hudgins, *op. cit.*
64. J. V. Gates, Brahman breeder, *History of Barzee Brahmans,* Poteet, Texas, 1978, p. 5.
65. *Ibid.*, p. 10.
66. J. T. White, Brahman breeder and former ABBA President, (interview), Hearne, Tx: Nov. 12, 1980.
67. *Ibid.*
68. Hudgins, *op. cit.*
69. Hammond, M. E., Manager of W. H. Stuart Ranch, Inc. (interview), Bartow, Fla., July 19, 1980.
70. Dr. A. O. Rhoad, Chief Animal Industry, Dept. Inter-American Institute of Agricultural Sciences, "New Blood and the American Brahmans," *Brahman Breeder-Feeder,* (Reprint), Houston: 1946.
71. Dr. W. S. Jacobs, "Morias Importation," ABBA Papers; Dr. W. S. Jacobs, "History of Brahman Importations," *American Brahman Journal,* Feb. 1949.
72. *Ibid.; Margaret Watkins, op. cit.*
73. *Ibid.*
74. Jacobs, "History of Brahman Importations," *op. cit.; Hudgins, op. cit.*
75. Dr. W. S. Jacobs (Letter to Gail Whitcomb, ABBA President), Aug. 10, 1948, ABBA Papers.
76. *Ibid.*, pp. 1-2.
77. Jacobs, "Morias Importation," *op. cit.*
78. Jacobs, (Letter to Gail Whitcomb), *op. cit.;* Milby B. Butler (Letter to Mrs. Frankye Thomas of Florida), Houston: *Breeder-Feeder,* Aug. 20, 1945.
79. Dr. W. S. Jacobs, *History Given ABBA Office,* Houston: 1949, pp. 5-6.
80. *Ibid.*, p. 4.
81. *Ibid., p. 8.*
82. Watkins, op. cit.
83. *Ibid.;* "Dr. W. S. Jacobs Shares Brahman Monopoly with Friends," *American Brahman Journal,* Houston: Sept., 1948, pp. 15-17. Dr. W. S. Jacobs, "The Expansion of 100% Zebu Blood," *The Zebu Journal,* Sept., 1949, p. 12.
84. ABBA Records, ABBA office, Houston.
85. "Historia de un hombre con historia," *Cebu,* Vol. 5, July-August, 1979, Indianapolis·4-50 piso, Mexico 18. D.F., p 61.
86. *Ibid.*, pp. 67-68.
87. *Ibid.*, Translation, ABBA Papers, p. 4.
88. "Esteban Garcia," (interview made by Wendell Schronk, ABBA, and transcribed), ABBA Papers, Houston.
89. *Ibid.*, p. 2.
90. *Ibid.*, p. 2.
91. *Ibid.*, pp. 2-3.
92. Edgar Hudgins, ABBA Director, Hungerford, Texas, July 16, 1980.
93. *Ibid.*, "Esteban Garcia," *op. cit.*, p. 3.
94. "Esteban and Eligio Garcia of McAllen, Texas," *Brahman Breeder-Feeder,* July, 1946, pp. 44-46.
95. James O. Sanders, Texas A&M, "History and Development of Zebu Cattle in the United States," *Journal of Animal Science,* Vol. 50, No. 6, 1980, pp. 1196-1199.
96. "Esteban Garcia," *op. cit.*
97. *Ibid.*, p. 6; Sanders, *op. cit.*, p. 1199.
98. Esteban Garcia, *op. cit.*, pp. 6-7.
99. Henry Clay Koontz, Brahman breeder, "Early Red Brahman History," *Red Brahman Directory,* 1976, pp. 28-29 and 53-55; Paul Horn, "The Reds From Brazil," *The Cattleman,* 1978, p. 44.
100. Koontz, *op. cit.*
101. Hudgins, *op. cit.*
102. Roger McKinley John, "A History of the Brahman Cattle Industry in Texas," Texas A&M, Thesis for work on Master of Arts, pp. 45-47.
103. E. C. "Buck" Sharman, USDA, interview, Hyattsville, Md., Feb. 8, 1980; Watkins, *op. cit.*, Dial, *op. cit.*, September 5, 1981.
104. ABBA Papers, Houston.
105. J. D. Sartwelle, *op. cit.*

Chapter V

1. Wendell Schronk, ABBA Exec. Vice-president (interview, ABBA office), Houston: Feb. 22, 1980; Henry Clay Koontz, ABBA Director and Brahman breeder (interview, HK Ranch), Placedo, Texas, Aug. 1, 1980.
2. Margaret Watkins, Brahman breeder and former Recording Secretary for ABBA, (interview, Edna, Texas), June 30, 1980.
3. Koontz, *op. cit.*
4. Henry Clay Koontz, "Early Red Brahman History in the United States," *Red Brahman Directory, 1976, p. 1; Roger McKinley John, A History of the Brahman Cattle Industry in Texas (a Thesis for completion of M.A. Degree, Texas A&M), College Station, 1966, p. 59.*

5. *Koontz, "Early Red Brahman History in the United States," op. cit.*, pp. 28-29; Harry Gayden, former ABBA Exec. Secretary (interview, Dallas), Oct. 4, 1980.

6. Steve Allen, Manager of Santa Elena Ranch, (interview, Madisonville, Tx.), Jan. 10, 1981.

7. Rodney Hammond, "Beginning of the Breed of Brahman Cattle in the United States," (interview Mr. Ernest Steinmann), *International Brahman Review;* "Rainbow S Ranch - El Rancho Pinto" W. C. Steinmann and J. Nyles Steinmann, *Red Brahman Directory;* ABBA Papers, Houston.

8. Koontz, "Early Red Brahman History . . . ." *op. cit.*, p. 28-29; "Stanley Kubela Prefers Red Brahmans," *The Zebu Journal,* Sept., 1950.

9. Koontz, "Early Red Brahman History. . . ." *op. cit.*, p. 28.

10. Koontz, *Ibid.*, p. 29.

11. Milby B. Butler, Brahman breeder of League City, Tx. (letter to Mrs. Frankye Thomas of Sarasota, Florida), Aug. 20, 1945, ABBA Papers, Houston; Dr. W. S. Jacobs, ABBA Director, (letter to Gail Whitcomb, ABBA President), Aug. 10, 1948, ABBA Papers, Houston.

12. Koontz, *op. cit.*, p. 29.

13. Butler, *op. cit.*, Koontz, *op. cit.*, p. 29.

14. Koontz, *op. cit.*, p. 53.

15. ABBA Minutes, 1924-1947, ABBA Office; Houston; Edgar Hudgins, ABBA Director and former ABBA President (interview, Hungerford, Tx.), July 8, 1980.

16. ABBA Minutes, 1924-1947, *op. cit.*

17. ABBA Papers, Houston, Texas.

18. Koontz, *op. cit.*, pp. 53-54.

19. *Ibid.*, p. 54; Paul Horn, "The Reds From Brazil," *The Cattleman,* 1978, p. 44; "Red Brahmans That Breed Red," *Zebu Journal,* Sept. 1950, p. 28.

20. Koontz, *op. cit.*, p. 55; "The Red Brahmans at Cherokee Ranch," *The Zebu Journal,* Sept., 1950, pp. 22-23.

21. Koontz, *op. cit.*, p. 55.

22. Koontz, interview, *op. cit.*

23. *Ibid.*

24. *Ibid.*

25. "The Red Brahmans at Cherokee Ranch," *op. cit.*, p. 22.

26. Henry Clay Koontz, "HK Ranch History," *Red Brahman Directory, p. 15.*

27. *Ibid.*

28. *Ibid.*

29. Joe Dial, ABBA Director, "History of Brahman Cattle Registered in ABBA," (transcript of speech), p. 6.

30. Paul W. Horn, "Tom Dean of Cherokee Ranch," *Zebu Journal,* July-August, 1980, pp. 28-30.

31. *Ibid.*, p. 28.

32. *Ibid.*, p. 29.

33. Bill Dutton, "Joint Red Brahman Sale Set at Sequin," *Farm and Ranch,* Mexia, Texas, May 5, 1978, pp. 10-11. (ABBA Papers, Houston).

34. "Red Brahmans on H. Cross Ranch," *The Zebu Journal,* Sept., 1950, p. 16.

35. "History of M.S. Holsworth Ranch," *Red Brahman Directory.* (ABBA Papers, Houston).

36. "History of McKellar Brahman Herd," Brochure published by McKellar Ranch (ABBA Papers, Houston).

37. *Ibid.;* Joe Ed Brockett, "The Brahman Cattle of Hacienda Pachonte," *The Brahman Journal,* Sagebrush Publ. Co., Sealy, Texas, July, 1980.

38. Henry Clay Koontz, interview, *op. cit.*

39. *Ibid.*

40. G. T. Stack, ABBA Director and Brahman breeder, interview, Feb. 10, 1981.

Chapter VI

1. Harry Gayden, former Executive Secretary of ABBA, (interview, Houston, Tx.), Feb. 22, 1980.

2. Lloyd Clyburn, unpublished manuscript on American Brahman, pp. 48-49. T. R. Havins, "Texas Fever," *op. cit.*, p. 161.

3. Clyburn, *op. cit.*, p. 49.

4. A. P. Borden (Letter to Hon. R. F. Broussard regarding Brahman cattle imported from India in 1906, Pierce, Texas), May 29, 1911 (copy property of ABBA, Houston).

5. Chris Emmett, *Shanghai Pierce, A Fair Likeness,* Norman: Univ. of Okla. Press, 1953, pp. 303-304.

6. Jay L. Lush, "Plan of the Experiments at the Texas Station," *Breeder Feeder,* Houston, Nov. 1941, p. 3-11.

7. Jay L. Lush, "Percentage of Calf Crop and Other Breeding Differences," *op. cit.*, p. 9.

8. Edgar Hudgins, ABBA Director and Brahman breeder, Hungerford, Tx., July 8, 1980, interview.

9. Marvin Koger, Livestock Geneticist and Professor of Animal Science, University of Fla., Gainesville: July 2, 1980, interview, J. L. Baughman, "Climate, Cattle and Crossbreeding . . . ." *Texas Journal of Science,* June 30, 1951, p. 263.

10. Joe Dial, ABBA Director and Brahman breeder, Placedo, Tx., Aug. 5, 1980, interview; Marvin Koger, *op. cit.*

11. J. W. Turner, *Genetic and Biological Aspects of Zebu Adaptability*, L.S.U., Baton Rouge: 1979, pp. 4-5; Jay L. Lush, *op. cit.*, p. 10; *The American Brahman*, ABBA publication, 1979, pp. 5-6.

12. Edgar Hudgins, ABBA Director and Brahman breeder, interview, July 18, 1980.

13. Clyburn, *op. cit.*, 76-77, pp. 103-107; J. L. Baughman, *op. cit.*, pp. 263-264.

14. Koger, *op. cit.*, interview, July 2, 1980

15. R. D. Randel, *Interrelationship of Endocrine and Physicological Events During the Estrous Cycle in Brahman Cattle*. Texas A&M, Overton: 1979; J. W. Turner, *op. cit.*, p. 8 (cited from J. R. Howes, *Crossbreeding Beef Cattle*, Univ. of Fla., 1963.)

16. Clyburn, *op. cit.*, pp. 14-16.

17. *The Brahman Handbook*, ABBA publication, Houston: 1979. p. 18. *Empty Tea Cup Fills Texas With Brahmans, op. cit.*, p. 14 (cited from USDA Farmers' Bulletin No. 1361).

18. "ABBA Materials," Houston; Leon Locke, ABBA President, *American Brahman Cattle and Their Influence on World Beef Production. (copy of speech made at ABBA Convention) Houston: 1966, p. 10.*

19. Hudgins, Walter T. Brahman Cattle, Hungerford, Texas.

20. Clyburn, *op. cit.*, pp. 18-22; *American Brahman Cattle*, ABBA publication, *op. cit.*, p. 9; *The American Brahman Breed*, ABBA publication, *op. cit.*, p. 6; Edgar Hudgins, *op. cit.*

21. Clyburn, *op. cit.*, p. 15.

22. *Ibid.*, pp. 15-16; C. M. Kincaid, *Breed Crosses with Beef Cattle in the South*, Southern Cooperative Series Bulletin 81 USDA, Georgia Coastal Plain Experiment Station, 1962, p. 9 and p. 22.

23. Locke, *op. cit.*, pp. 11-16; Kirby Cunningham, ABBA, "Fertility in the Brahman Cow," *The Cattleman*, Feb., 1971, p. 1; "Performing Ability," *American Brahman Cattle*, ABBA publication, *op. cit.*, pp. 8-9.

24. Cunningham, "Fertility in the Brahman Cattle," *op. cit.*, pp. 1 and 2.

25. Cartwright and Fitzhugh, *Selecting and Breeding the Unique Brahman, op. cit.*, Part IV.

26. T. C. Cartwright, Texas A&M, "Longevity, A Valuable Fertility Trait," *The Cattleman*, Feb., 1971, p. 1.

27. Clyburn, *op. cit.*, p. 13.

28. Gayden, *op. cit.*, (interview); Clyburn, *op. cit.*, p. 62.

29. Clyburn, *op. cit.*, pp. 62-63; Harry Gayden, *op. cit.*

30. Clyburn, *op. cit.*, p. 63.

31. R. L. Moore, H. W. Essig and L. J. Smithson, "Influence of Breeds of Beef Cattle on Ration Utilization," *Journal of Animal Science*, 1975, p. 203.

32. Clyburn, *op. cit.*

33. J. W. Sartwelle, *Interesting Points About Brahman Cattle* (transcript of radio address, KPRC, Houston, Oct. 14, 1939).

34. Gayden, *op. cit.*, (interview).

35. ABBA Historical Materials, op. cit.

36. C. I. Bray, Louisiana Agricultural Experiment Station, "L.S.U. Professor Tells of Experimental Work Done with Crossbreds; . . . ." *The Florida Cowman, Feb. 1952, p. 66; W. C. McCormick and B. L. Southwell, "A Comparison of Brahman Crossbred with British Crossbred Cattle, Journal of Animal Science*, Vol. 16, Georgia Coastal Plain Experiment Station, Tifton: Feb. 1957, pp. 215-216.

37. *American Brahman Cattle*, ABBA publication, *op. cit.*, pp. 10-11; F. M. Peacock, M. Koger, W. G. Kirk, E. M. Hodges, and A. C. Warnick "Comparative Performances of Various Beef Breeds and Crosses in Florida," *Crossbreeding Beef Cattle, op. cit.*, pp. 43-48.

38. J. E. Lawson, Animal Geneticist, "Brahman X Hereford Hybrids Excel in Southern Alberta," *Cattleman*, p. 8.

39. Hudgins, *op. cit.*

40. "Brahmans, Australian History," *Diary For Cattleman*, Australian Brahman Breeders' Association; Rockhampton, Queensland, Australia, 1971, pp. 2-3.

41. *Ibid.*, p. 6.

42. *Ibid.*, pp. 3-5; ABBA Materials, *op. cit.*

43. Hudgins, *op. cit.*, ABBA Materials, *op. cit.*; R. B. Kelley (Progress Reports Nos. 2, 3, and 4 of the Council for Scientific and Industrial Research, Commonwealth of Australia), *Zebu* (Brahman) *Cross Cattle and Their Possibilities in North Australia*. Sydney, 1936, 1938, 1940; Clyburn, *op. cit.*, pp. 66-67.

44. Gayden, *op. cit.*

45. Clyburn, *op. cit.*, p. 67.

46. J. W. Turner, *Genetic and Biological Aspects of Zebu Adaptability*, L.S.U., Baton Rouge: 1979, (copy of speech presented at symposium), p. 4.

47. *Ibid.*, p. 5.

48. *Ibid.*, p. 6.

49. Clyburn, *op. cit.*, p. 11; Lush, "Brahman Cattle, 1920-1929," *op. cit.*, p. 11.
50. Study made by Kelley cited by ABBA Director Wayne Fisher, *United Nations in Agreement on One Subject - Livestock*, ABBA Materials, Aug. 2, 1948, p. 10.
51. *Ibid.*, p. 10.
52. Turner, L.S.U., *op. cit.*, (interview), June 27, 1980.
53. Koger, U.F., *op. cit.*, (interview), July 2, 1980.
54. Quote by W. D. Hunter cited by J. W. Sartwelle in letter to M. K. Reddy, Cornell University, Nov. 17, 1924, ABBA Materials, p. 6.
55. J. W. Sartwelle, "Brahmans Have Interesting, Unusual Background . . . ." *Coastal Cattleman*, Beaumont: Dec. 1939, p. 5-6; ABBA Materials, *op. cit.*
56. Turner, *op. cit.*
57. J. W. Sartwelle, (Letter to M. K. Reddy, Cornell Univ.), Nov. 17, 1924.
58. Turner, *op. cit.*
59. J. W. Turner, *Genetic and Biological Aspects of Zebu Adaptability*, *op. cit.*, (cited from J. R. Howes, "Blood Composition and Physiology of Brahman and Herefords," *Crossbreeding Beef Cattle*. Univ. of Fla. Press, Gainesville). p. 4.
60. Leon Locke, *American Brahman Cattle and Their Influence on World Beef Production*, speech made at ABBA convention, 1966. Albert O. Rhoad, "The Iberia Heat Tolerance Test" *Tropical Agriculture*, Vol. 21: 1944, pp. 162-164.
61. T. E. Allen, 1962. "Responses of Zebu, Jersey and Zebu X Jersey Crossbred Heifers to Rising Temperatures . . . ." *Australian Journal of Agriculture*, Res. 13: 165 (cited by J. W. Turner in *Genetic Aspects of Zebu Adaptability*, *op. cit.*, pp. 4-5.
62. T. C. Cartwright, "Responses of Beef Cattle to High Ambient Temperatures," *Journal of Animal Science*, 14: 35, 1955 (cited by J. W. Turner in *Genetic Aspects of Zebu Adaptability*, *op. cit.*, p. 5.
63. Rhoad, *op. cit.*
64. *op. cit.*, Clyburn, *op. cit.*, pp. 4-5; 21.
65. Clyburn, *op. cit.*, pp. 21-22.
66. Clyburn, *op. cit.*, pp. 67-68.
67. Clyburn, *op. city.*, p. 68.
68. Clyburn, *op. cit.*, p. 25.
69. Rhoad, *op. cit.*, *The Brahman Hybrid*, ABBA publication, Houston, p. 13.
70. ABBA Materials, *op. cit.*
71. *The Brahman Hybrid*, *op. cit.*, p. 17.
72. R. E. Patterson, Texas A&M, *Hybrid Vigor in Brahman-Hereford Crosses*, May 1964; Clyburn, *op. cit.*, pp. 53, 66-67; "Brahmans, Fad, Fancy or Fact," *op. cit.*, p. 11.
73. Clyburn, *op. cit.*, p. 68; J. L. Baughman, *op. cit.*, p. 271.
74. Dr. Harold J. Brooks, "The Brahman Contribution to Commercial Dairying," *Proceedings of The First Brahman Congress*. Southeastern Brahman Breeders' Assoc., Sarasota: 1947, pp. 70-89.
75. *Ibid.*, p. 74.
76. *Ibid.*, p. 75.
77. J. L. Baughman, *op. cit.*, pp. 285-286.
78. *The Brahman Hybrid*, *op. cit.*, pp. 7-8; Leon Locke, *op. cit.*, p. 11.
79. J. L. Baughman, *op. cit.*, pp. 286-287.
80. R. W. Phillips, *Breeding Livestock Unfavorable*. F.A.O. Agricutural Studies: 1948, pp. 1-182; R. W. Phillips, "Adaptability of Cattle to Tropical and Subtropical Climates," *Cattleman*, Texas and Southwestern Cattle Raisers Assoc., Fort Worth: 1946, Vol. 33; Wayne Fisher, *United Nations in Agreement on One Subject . . . . op. cit.*, pp. 5-7.
81. J. L. Baughman, *op. cit.*, pp. 281-288; John Rouse, *World Cattle*, Vols. I and II, University of Oklahoma Press, Norman: 1970.
82. Letter from J. W. Sartwelle to D. M. Reddy, *op. cit.*; N. R. Joshi, and Ralph Phillips, *Zebu Cattle of India and Pakistan* FAO Study, Rome: 1953, pp. 153-166.
83. Hudgins, *op. cit.*
84. Clyburn, *op. cit.*, pp. 72-76.
85. *Ibid.*, p. 77.
86. *Ibid.*, pp. 77-78.

Chapter VII

1. Leon Locke, former ABBA President and ABBA Director, *American Brahman and Their Influence on World Beef Production*, ABBA Papers, Houston: 1966, p. 10.
2. Edgar Hudgins, ABBA Director, *Guzerat Brahman Cattle and Facts on Crossbreeding*, ABBA Papers, Hungerford, Texas, 1934.
3. William Broyles, Jr., "The Last Empire," *Texas Monthly*, Oct., 1980, p. 28.
4. Black, W. H., "Brahman Cattle in America," *American Feed and Grain Dealer*, pp. 10, 11, 34, 35, 42, Nov. 1946.

5. Edgar Hudgins, ABBA Director, unpublished manuscript, Hungerford: 1980.
6. Don Wideman, "First American Beef Breed," *Stockman Farmer*, 1978, pp. 26-28.
7. ABBA materials, Houston.
8. J. L. Baughman, "Climate, Cattle and Crossbreeding . . . ." *The Texas Journal of Science*, June, 1951, p. 274.
9. Baughman, *op. cit.*, pp. 275-276.
10. *Ibid.*
11. Baughman, *op. cit.*, pp. 269-270.
12. David Whitman, "Breeding Up to Brangus With $1/4$ Bloods and $1/2$ Bloods," *Brangus Journal*, May: 1977, p. 20.
13. *Ibid.*
14. *Ibid.*
15. Roy W. Lilley, Ex. Sec., International Brangus Breeders Ass'n., ABBA Papers, Houston.
16. Edgar Hudgins, *op. cit.*
17. *Beef Cattle in Florida*, Bulletin No. 28, Fla. Dept. of Agriculture and Consumer Services, Tallahassee: 1976, pp. 26-29.
18. *World Book Encyclopedia*, Vol. 3, 1965; J. L. Baughman, *op. cit.*, pp. 270-271.
19. *Beef Cattle in Florida*, *op. cit.*, pp. 29-30.
20. J. L. Baughman, *op. cit.*, pp. 270-271; "Braford Cattle Rapidly Coming to Forefront in Coastal Livestock Picture," *Coastal Cattleman*, Beaumont: June, 1940; pp. 5-6.
21. Wendell Schronk, Executive Vice-President of ABBA, interview; Aug. 30, 1980.
22. J. L. Baughman, *op. cit.*, pp. 273-274; *Beef Cattle in Florida*, *op. cit.*, pp. 30-31.
23. J. L. Baughman, *op. cit.*
24. Marvin Koger, Professor of Animal Breeding and Genetics, *Characteristics of Types and Breeds of Cattle in the Tropics*, Univ. of Fla., Gainesville: 1980, p. 14.
25. *Ibid.*
26. J. L. Baughman, *op. cit.*, p. 274.
27. John Walter, "What Is a Simbrah?" *Simmental Shield;* Bozeman: American Simmental Assoc., July, 1980, pp. 59-60.
28. *Southern Beef Producer*, Nashville, Tenn.
29. Edgar Hudgins, *op. cit.*; *The Brahman Hybrid*, ABBA publication, Houston: 1978, p. 10.
30. Koger, *op. cit.*, p. 15.
31. *Ibid.*, p. 17.
32. *Ibid.*, p. 18.
33. *The Brahman Hybrid*, *op. cit.* J. W. Turner, "Genetic and Biological Aspects of Zebu Adaptability," *Journal of Animal Science*, Vol. 50, No. 6, 1980, p. 1204; Wendell Schronk, *op. cit.*; Edgar Hudgins, *op. cit.*
34. Ronald Smoak, Professor of Agriculture, NFJC; interview, Nov. 5, 1980.
36. Wendell Schronk, *op. cit.*; *World Book Encyclopedia*, Vol. 3, Field Enterprises, Chicago: 1965, p. 236.
37. Harry Gayden, *op. cit.*
38. *The Brahman Hybrid*, ABBA publication, Houston: 1978, p. 9 (cited from 1950-1964 studies performed at McGregor, Tx.)
39. *Ibid.*
40. *Ibid.*
41. ABBA Minutes, 1978-1980; Marvin Koger, *op. cit.*, (interview).
42. ABBA Minutes, 1978-1980.
43. *The Brahman Hybrid*, *op. cit.* (cited from 1962-1966 studies at Angleton, Tx.), p. 7.
44. *Ibid.*, p. 13.
45. ABBA Minutes, 1978-1980; ABBA Ad, *Brahman Journal*, *op. cit.*
46. Edgar Hudgins, *op. cit.*; Wendell Schronk, op. cit.
47. *American Brahman Cattle*, published by ABBA, Houston: 1979, pp. 12-14; ABBA Handout on F_1 cow, *op. cit.*
48. Clyburn, *op. cit.*, pp. 24-26; Zerle Carpenter. "Effects of Crossbreeding on Carcass Characteristics," *Crossbreeding Beef Cattle, Series 2*, Edited by Marvin Koger, Tony J. Cunha, Alvin C. Warnick, Univ. of Fla. Press, Gainesville: 1973, pp. 163-182; Dr. Frank Orts, Texas A&M. "Carcass Characteristics of Brahman Crosses," *The Cattleman*, February, 1971, (Reprint).
49. *Ibid.*
50. Wendell Schronk, *op. cit.*
51. Joe Ed Brockett, "Brahman-Hereford Field Day," *The Brahman Journal*, July, 1980, p. 24.
52. W. C. "Cal" Burns, Director USDA Beef Cattle Research Center, Speech given at Brahman-Angus Field Day, Brooksville, Fla., July 18, 1980.
53. Margaret Watkins, tormer Recording Secretary ABBA, (interview) Edna, Texas; July 19, 1980.
54. Wendell Schronk, *op. cit.*
55. Vance Ehmke, "Brahman Crosses for the Great Plains," *The Brahman Journal*, *op. cit.*, pp. 8-10.
56. Wendell Schronk, "Bill Cowan, a Man as Big as the Country . . ." *The Brahman Journal*, *op. cit.*
57. *Ibid.*

58. Bob Bracher, "Brahman Bulls on Angus Cows," *Cattleman Magazine*, Fort Worth: Feb., 1972 (Reprint).
59. Paul W. Horn, "Black Hybrids," *Cattleman Magazine*, Fort Worth: Feb., 1973 (Reprint).
60. *Ibid.*
61. Joe Ed Brockett, "Forty Years of Brahman Cattle in Alabama," *The Brahman Journal*, *op. cit.*, Nov., 1980, pp. 8-9, 38.
62. Joe Ed Brockett, "Running Brahman Crosses," *The Brahman Journal*, *op. cit.*, Aug., 1980, pp. 4-5.
63. *The Florida Cattleman*, Kissimmee.
64. *Ibid.*
65. Joe Ed Brockett, "The American Brahman," *The Brahman Journal*, *op. cit.* May, 1980, p. 10; Marvin Koger, interview, *op. cit.*; *Wendell Schronk, op. cit.*; *Roman L. Hruska, U.S. Meat Research Center, Progress Report No. 7.*
66. Leon Locke, American Brahman Cattle and Their Influence on World Beef Production, speech, ABBA meeting, Houston: 1966, pp. 10-11.
67. Mike Mendenhall, Colorado cattleman, quote from ad in The Brahman Journal.
68. Harry Gayden, *op. cit.*
69. Jim Park, "Henry O. Partin, Pioneer Cattleman," *American Brahman*, Sept. 1950, pp. 11-12; "Cattle Kingdom of Kissimmee," *Tractor Farming*, July-August, 1952, p. 12.
70. Joe Ed Brockett, "From Out of the Swamp," *The Brahman Journal*, *op. cit.*, Feb. 1980, pp. 8-10.
71. D. E. Franke, "Breed and Heterosis Effects of American Zebu Cattle," *Journal of Animal Science*, Vol. 50, No. 6, 1980, p. 1210. Marvin Koger, "Effective Crossbreeding Systems Utilizing Zebu Cattle," *Journal of Animal Science*, *op. cit.*, p. 1219; Marvin Koger, *op. cit.*, interview.
72. "Cal," W. C. Burns, *op. cit.*
73. Wendell Schronk, *op. cit.*; Edgar Hudgins, *op. cit.*
74. Clyburn, *op. cit.*

Chapter VIII

1. Edgar Hudgins, ABBA Director and Brahman breeder (interview at Hungerford, Texas), July 19, 1980.
2. ABBA Minutes, 1924-1947, ABBA Headquarters, Houston, Texas.
3. ABBA Minutes, 1973-1980, *op. cit.*
4. *Ibid.*
5. *Ibid.*; ABBA Transfer Records, ABBA Headquarters, Houston, Texas.
6. Harry Gayden, former ABBA Executive Director (interview at Dallas, Texas), Oct. 11, 1980.
7. ABBA Minutes, 1972, *op. cit.*
8. ABBA Minutes, 1968-1980, *op. cit.* Wendell Schronk, ABBA Executive Vice-President, (interview at Houston, Texas), April 4, 1981.
9. Schronk, *op. cit.*
10. ABBA Minutes, 1924-1947, *op. cit.*
11. ABBA Minutes, 1976-1980, *op. cit.*
12. ABBA Minutes, 1964-1971, *op. cit.*; Joe Dial, ABBA Director and Brahman breeder (interview at Placedo, Texas), June 7, 1981.
13. "Register of Renown," *The Brahman Journal*, Sagebrush Publishing Co., Inc., Eddy, Texas, Aug. 1981.
14. Schronk, *op. cit.*
15. "All American Draws 285 Entries from Four States," *The Brahman Journal*, *op. cit.*, Sept., 1980.
16. ABBA Minutes, 1972, *op. cit.*
17. ABBA Minutes, 1963, 1975-1980, *op. cit.*
18. ABBA Minutes, 1979-1980, *op. cit.*; Schronk, *op. cit.*
19. ABBA Minutes, 1980; Schronk, *op. cit.*
20. ABBA Minutes, 1977, *op. cit.*
21. ABBA Minutes, 1974, *op. cit.*
22. ABBA Minutes, 1947-1980.
23. *Ibid.*; Schronk, *op. cit.*; Dial, *op. cit.*; G. T. Stack, Brahman breeder and exporter (interview at Tampa, Florida), Feb. 3, 1981.
24. J. D. Sartwelle, ABBA Director and Brahman breeder (interview at Houston, Texas), Oct. 18, 1981. Wendell Schronk, ABBA Executive Vice President (Notes made by Schronk on trip to Zimbabwe, Africa, Aug. 1980).
25. ABBA Minutes, 1978-1980, *op. cit.*
26. *Ibid.*
27. Wendell Schronk, ABBA Executive Vice-President, (interview) *op. cit.*, August 12, 1981.
28. "ABBA Headquarters at Six Locations Since 1924," *The Brahman Journal (Fiftieth ABBA Anniversary Issue)*, *op. cit.*, Feb., 1974.
30. *Ibid.*
31. Schronk, *op. cit.*
32. ABBA Minutes, *op. cit.*, 1924.

Chapter IX

1. C. Clausen Steampship Company, Ltd., (informational and advertising brochure issued by C. Clausen Co.); Claude Dobbins, former livestock marketing specialist U.S.D.A. and presently advisor to C. Clausen Steampship Co. (interview at Alexandria, Va.), July 30, 1980.

2. "World War II," *Encyclopedia Britannica*, "W" Volume, 1976; *Port City Stockyards, 1931-1956* (Short resume of Port City Stockyards' first 25 years), published by Port City Stockyards, Houston: 1956, p. 10.

3. Byron Johnson, former purebred herd manager of Norris Cattle Co. (interview at Ocala, Florida), Nov. 5, 1981.

4. *Ibid.*; G. T. Stack, Brahman breeder and exporter (interview at Tampa, Florida), Nov. 6, 1981.

5. *Port City Stockyards*, *op. cit.*, p. 10.

6. *Ibid.*, pp. 10-11.

7. J. D. Sartwelle, ABBA Director and Brahman breeder (interview at Houston, Texas), Nov. 3, 1981.

8. George O. Puig, former President of Houston Freight Forwarding Co. (interview at Houston, Texas), Nov. 4, 1981.

9. Margaret Watkins, Brahman breeder and former ABBA Recording Secretary (interview at Edna, Texas), Nov. 5, 1981.

10. Stack, *op. cit.*, July 15, 1980.

11. *Ibid.*, Nov. 3, 1981.

12. Puig, *op. cit.*

13. *Ibid.*

14. Sartwelle, *op. cit.*; Stack, *op. cit.*; Watkins, *op. cit.*

15. Stack, *op. cit.*; "Panamanians Buy Cattle," Fla. Department of Agriculture and Consumer Services News Release, Tallahassee: Sept., 1980.

16. Watkins, *op. cit.*

Chapter X

1. "Historia de Un Hombre con Historia," *CEBU*. Vol. 5. Piso, Mexico, 18. D.F: July-August, 1979. p. 1.

2. *Ibid.*, p. 2.

3. Edgar Hudgins, ABBA Director, J.D. Hudgins, Inc. Hungerford, Texas: Juy 14, 1980.

4. *Ibid.*, June 5, 1980.

5. "Jacobs Bull to Honduras," *Coastal Cattleman*, Beaumont: August, 1939, p. 14.

6. "ABBA Transfer Records, 1939-1980," American Brahman Breeders Association, Houston, Texas.

7. "Duffy" Marceaux, Brahman breeder, Kaplan, La., July 30, 1980; J. D. "Bubba" Sartwelle, Jr. Sartwelle Brahman Ranch Ltd. Houston: June 12, 1981.

8. Marceaux, *Ibid.*

9. Margie Forbes Rogillio, Brahman breeder, Greenwood Springs, Louisiana: July 31, 1980.

10. "ABBA Minutes, 1924-1980," American Brahman Breeders Association, Houston, Texas.

11. Hudgins, *op. cit.*, February 20, 1980.

12. "Florida," *Brahman Breeder-Feeder*, Houston: August, 1946, p. 38.

13. *Ibid.*, pp. 38-39, p. 52.

14. *Ibid.*, p. 63; Margaret Watkins, former ABBA Recording Secretary, (interview) Edna, Texas. July 18, 1980.

15. Herman Beville, Florida Brahman breeder and former ABBA Director, (interview), Bushnell, Florida: July 30, 1980.

16. Wayne Fisher, "Brahmans in Hawaii," *The Zebu Journal.* Cotulla, Texas: Sept., 1950; "ABBA Minutes," *op. cit.*, "Brahman Cattle in Arizona," *Coastal Cattleman*. Beaumont: March, 1941.

17. "Brahman Cattle are Getting Around in the Far West," *Brahman Breeder-Feeder*. July, 1946.

18. Margaret Watkins, former ABBA Recording Secretary, Edna, Texas: June 30, 1980; Avon Davis, Brahman breeder, A. D. Ranch, Chowchilla, Calif.: Aug. 5, 1980.

19. Davis, *op. cit.*

20. Wendell Schronk, ABBA Executive Vice-President, ABBA, Houston: April 15, 1981.

21. Joe Dial, ABBA Director, Placedo, Texas, Sept. 4, 1981; Hudgins, *op. cit.*, Feb., 1981.

22. "Southwest Bull Classic," *Brahman Journal*, Sagebrush Publishing Co., Sealy, Texas, May, 1980.

23. *Ibid.*

24. Charles Rodeffer, "Brahman Blood Brought Here by Cattlemen," Norfolk, Virginia newspaper, Norfolk: June 12, 1949.

25. Bill Zipf, "Brahmans Were Most Popular Livestock Display at Fair," *The Columbus Dispatch*, Columbus, Ohio. Sept. 5, 1951.

26. Lewis Patterson, Brahman breeder, (interview), Concord, North Carolina, July 25, 1981.

27. Joe Akerman, *The Florida Cowman, A History of Florida's Cattle Industry*, FCA,

Kissimmee: 1976; John E. Rouse, *The Criollo, Spanish Cattle in the Americas, University of Oklahoma Press: Norman: 1977, p. 3-21.*

28. Rouse, Ibid., pp. 4, 34; James O. Sanders, Professor of Animal Science (Texas A&M), "A History of the Zebu in Brazil," (speech), College Station, Texas. 1980.

29. Rouse, *op. cit.,* p. 93.

30. "ABBA Minutes, 1924-1980," *op. cit.*

31. "Brahman Cattle in Cuba," *Brahman Breeder-Feeder*, Houston: May, 1947, pp. 51-56.

32. Hudgins, *op. cit.*, July 1, 1981.

33. Lionel Chambers, "Central Romana," *The Brahman Journal* (Special 50th Anniversary Issue), Chaparral Publishing Co., Inc., Sealy, Texas, Feb., 1974.

34. "ABBA Minutes, 1924-1980," *op. cit.* "Martinique," *Encyclopedia Britannica*, "M" Volume, New York: 1976.

35. G. T. Stack, ABBA Director (former ABBA President) and exporter, Tampa: Aug. 4, 1980; Feb. 10, 1980.

36. "Venezuela" and "Columbia," *World Book Encyclopedia*, "V" and "C" volumes, Chicago: 1968. Rouse, *op. cit.*, Univ. of Oklahoma Press, Norman: pp. 124-125.

37. Chaffee, F.H., Erichson, E.E.; Fortenbaugh, S.G.; Malihs, S.; and Stodter, J.H., *Area Handbook for Venzuela*. The American University, Washington, D.C., 1971, pp. 323-325. "U.S. Cattle to Improve South American Herds," *Foreign Agriculture*, Washington, D.C., March 1955, pp. 53-56; Rouse, *World Cattle*, Vol. I, University of Okla. Press, Norman: 1970, pp. 447-459.

38. "U.S. Cattle to Improve South American Herds," *op. cit.*; Joe Dial, *op. cit.*

39. "Columbia," *World Book Encyclopedia*, *op. cit.*; Rouse, *op. cit.*, p. 409.

40. "ABBA Transfer Records, 1939-1980," *op. cit*

41. *Ibid.*; Harry Gayden, former Executive-Secretary ABBA, Houston: August 5, 1981.

42. Stack, *op. cit.*

43. Rouse, *op. cit.*, pp. 410-412.

44. *Area Handbook for Venezuela*, *op. cit.*, Rouse, *op. cit.*, pp. 412-413.

45. Rouse, *op. cit.*, pp. 415-416.

46. Rouse, *The Criollo, Spanish Cattle in the Americas*, *op. cit.*, pp. 43, 66-71.

47. "Argentina," *Encyclopedia Britannica*, Vol. I, New York: 1976, Rouse, *op. cit.*, pp. 95-96.

48. "ABBA Transfer Records, 1939-1980," *op. cit.*: Dial, *op. cit.*

49. Koger, M.; Cunha, T.J.; Warnick, A.C., *Crossbreeding Beef Cattle, Series 2*. University of Florida Press, Gainesville: 1973, pp. 422-433.

50. Rouse, *World Cattle*, Vol. I, *op. cit.*, pp. 357-366.

51. "Cattle Populations of the World," U.S. Department of Commerce, Bureau of the Census; U.S.D.A.; Washington: 1970.

52. Schronk, *op. cit.* and Hudgins, *op. cit.*

53. Dr. Carlos Smith, Rural Society of the Triangulo Mineiro, "Trio of Brazilians Here For Fat Stock Show," *The Houston Post*, Houston: 1946; Rouse, *World Cattle, Vol. I, op. cit.*, pp. 383-388.

54. Rouse, *op. cit.*, p. 389.

55. "Cattle Populations of the World," *op. cit.*, Rouse, *op. cit.*

56. Hudgins, *op. cit.*, and Schronk, *op. cit.*

57. "Gran Chaco," *World Book Encyclopedia*, Vol. VIII, 1976; *Area Handbook of Paraguay*, *op. cit.*, pp. 197-200; Rouse, *op. cit.*, p. 427.

58. "ABBA Transfer Records, 1939-1980," *op. cit.*

59. Rouse, *op. cit.*, p. 428.

60. Frederick Lege III, FAS/USDA, "Ecuador," *Guide for U. S. Cattle Exporters*, Agriculture Handbook No. 217. USDA/FAS, Washington: 1976, p. 34; "ABBA Minutes, 1924, 1980," *op. cit.*; "ABBA Transfer Records, 1939-1980," *op. cit.*

61. "ABBA Minutes," *op. cit.*

62. "ABBA Transfer Records," *op. cit.*; G. T. Stack, *op. cit.*

63. Watkins, *op. cit.*

64. *Crossbreeding Beef Cattle, Series 2*, *op. cit.*, pp. 410-411; 419.

65. "ABBA Transfer Records," *op. cit.*

66. Hudgins, *op. cit.*, July 30, 1981.

67. Stack, *op. cit.*, July 30, 1981.

68. John Stiles, Director of Marketing, Fla. Dept. of Agriculture and Consumer Services, Tallahassee: Aug. 30, 1980.

69. "Panamanians Buy Cattle," News Release, Fla. Dept. of Agriculture and Consumer Services. Tallahassee: Sept., 1980.

70. "ABBA Export Records," *op. cit.*

71. Joe Ed Brockett, "The Brahman Cattle of Hacienda Pachonte," *The Brahman Journal*, *op. cit.*, July, 1980, p. 18.

72. *Ibid.*, p. 21; p. 105.

73. Watkins, *op. cit.*, Sept. 4, 1981; "Honduras," *World Book Encyclopedia*, Vol. VII, Chicago: 1976.

74. "ABBA Transfer Records, 1939-1980," *op. cit.*

75. Watkins, *op. cit.*

76. Stack, *op. cit.*

77. "ABBA Transfer Records, 1939-1980," *op. cit.*

78. Dial, *op. cit.*

79. Rouse, *World Cattle*, Vol. II, *op. cit.*, pp. 713, 751, 725, 735, 535, 763 and 625.

80. *Ibid.*, p. 743.

81. "Africa," *Encyclopedia Britannica*, Vol. I; David Cohen, Professor of African Studies, Johns Hopkins University, Baltimore: 1970.

82. *Ibid.*; Rouse, *op. cit.*, pp. 492, 493.

83. Rouse, *op. cit.*, pp. 636-637.

84. Gayden, *op. cit.*, October, 1980.

85. "ABBA Transfer Records, 1939-1980," *op. cit.*

86. Schronk, *op. cit.*, Aug. 5, 7, 1980.

87. *Ibid.*, *Hudgins, op. cit.*

88. Monty Banks, General Manager, McKellar Ranch, Mt. Pleasant, Texas, Aug. 1, 1980; "Zimbabwe Expo Features Brahman Show and Clinic," *Brahman Journal*, *op. cit.*, July, 1981, p. 32.

89. Rouse, *op. cit.*, p. 584.

90. Schronk, *op. cit.*

91. *Ibid.*

92. *Ibid.*

93. "Mozambique," *Encyclopedia Britannica; "ABBA Transfer Records, 1939-1980*, *op. cit.*, Rouse, *op. cit.*, pp. 639-648.

94. "Botswana," *World Book Encyclopedia*, Vol. II, *op. cit.*; *"Botswana," Encyclopedia Britannica, Vol. II, op. cit.* USDA, Statistical Reporting Service, Washington. 1976.

95. Gayden, *op. cit.*, Hudgins, *op. cit.*

96. ABBA Transfer Records, *op. cit.*, Rouse, *World Cattle*, Vol. II., *op. cit.*; *Encyclopedia Britannica*, Vol. I, *op. cit.*

97. ABBA Transfer Records, *op. cit.*

98. Gayden, *op. cit.*

99. Stack, *op. cit.*

100. Leslie "Bubba" Hudgins, Brahman breeder, J. D. Hudgins, Inc., Hungerford, Texas, April 12, 1980.

101. Schronk, *op. cit.*

102. Rouse, *World Cattle*, Vol II, *op. cit.*, pp. 681-684.

103. Arthur Oliver, Col. Ret. (British Army) *A Brief Survey of the Important Breeds of Cattle in India*, Govnt. of India Press, New Delhi: 1938, pp. 1-8.

104. Stack, *op. cit.*

105. P. N. Joshi, Supt. Darvan Dairy, Bhavnagar, India (Letter to Walter Hudgins, Hungerford, Texas, 1938), ABBA Papers, Houston.

106. *Encyclopedia Britannica*, "P" Volume, *op. cit.*; Rouse, *World Cattle*, Vol. II, *op. cit.*, p. 1014.

107. Stack, *op. cit.*

108. Stack, *op. cit.*; Dial, *op. cit.*

109. *USDA Guide for U. S. Cattle Exporters*, *op. cit.*, p. 111.

110. Stack, *op. cit.*

111. ABBA Transfer Records, *op. cit.*; Rouse, *World Cattle*, Vol. II, *op. cit.*, p. 955.

112. Marceaux, *op. cit.*, *World Book Encyclopedia*, Vol. 18, *op. cit.*

113. ABBA Transfer Records, *op. cit.*; Rouse, *World Cattle*, Vol. II, *op. cit.*, p. 935; pp. 941-942.

114. *USDA Guide for U. S. Cattle Exporters*, *op. cit.*, p. 94.

115. *Ibid.*, (quote from "one official").

116. ABBA Minutes, 1971-1980, *op. cit.*

117. Stack, *op. cit.*

118.-119. ABBA Transfer Records, *op. cit.*, Dial, *op. cit.*

120. Rouse, *World Cattle*, Vol. II, *op. cit.*, p. 1020.

121. Hudgins, *op. cit.*

122. Dial, *op. cit.*; Hudgins, *op. cit.*, ABBA Transfer Records, *op. cit.*

123. Hudgins, *op. cit.*; *USDA Guide for U. S. Exporters*, *op. cit.*

124. *Ibid.*, ABBA Transfer Records; Stack, *op. cit.*

125. James R. Simpson, "Exporting Florida Beef Breeding Cattle; Potential, Prospects, Problems," *The Florida Cattleman*, FCA, Kissimmee: Aug. 1980, pp. 44-45.

126. Kirby Cunningham, former Director of Public Relations of ABBA, "A Historic Herd for the Future," *The Cattleman*, Feb., 1971.

127. "Consumer and Market Research on Beef," *Beef Business Bulletin*, NCA, Jan. 16, 1981, p. 82.

128. "ABBA Minutes, 1924-1980," *op. cit.*

129. Stewart Fowler, "The Original Exotic and Beef Revolution in the South," *Cattleman Magazine*. Fort Worth: Feb., 1973.

130. Joe Ed Brockett, "From Out of the Swamp." *op. cit.*, Feb., 1980.

131. *Ibid.*, "A Look at Guatamala's Next Generation, Luis Morales," *The Brahman Journal*, *op. cit.*, April, 1980.

132. Roger McKinley John, *A History of the Brahman Cattle Industry in Texas*, a thesis for a Master of Arts Degree, Texas A&M, College Station: May, 1966, pp. 103-104.

133. *Ibid.*, p. 104.

134. Joe Ed Brockett, "First Brahman Certified Meat Sire," *op. cit.*, May 1, 1979, pp. 10-13.

AMERICAN BRAHMAN BREEDERS ASSOCIATION REGISTRATIONS, 1924-1981

Year	Registrations U.S.	Registrations Fg.	Registrations Total	Cumulative Registrations
1924-1933			2,765	2,765
1934-Mid 1936			1,868	4,633
Mid 1936-1938			12,029	16,662
1938				
1939			2,025	18,687
1940			4,409	23,096
1941			4,132	27,228
1942			5,795	33,023
1943			8,536	41,559
1944			7,054	48,613
1945			10,856	59,469
1946			10,623	70,092
1947	14,184	744	14,928	85,020
1948	14,170	975	15,145	100,165
1949	15,630	1,590	17,220	117,385
1950	17,733	1,444	19,177	136,562
1951	18,987	1,623	20,610	157,172
1952	22,312	1,701	24,013	181,185
1953	16,110	1,763	17,873	199,058
1954	15,267	1,743	17,010	216,068
1955	15,085	1,479	16,564	232,632
1956	12,437	2,303	14,740	247,372
1957	12,836	2,440	15,276	262,648
1958	11,573	2,473	14,046	276,694
1959	11,092	1,801	12,893	289,587
1960	12,082	933	13,015	302,602
1961	10,324	1,103	11,427	314,029
1962	10,437	1,724	12,161	326,190
1963	12,382	1,409	13,791	339,981
1964	11,735	2,438	14,173	354,154
1965	12,894	2,722	15,616	369,770
1966	12,134	3,147	15,281	385,051
1967	14,226	3,512	17,738	402,789
1968	11,911	3,188	15,099	417,888
1969	14,489	3,389	17,878	435,766
1970	14,995	3,224	18,219	453,985
1971	15,725	2,763	18,488	472,473
1972	19,325	3,189	22,514	494,987
1973	19,868	2,882	22,750	517,737
1974	21,215	4,051	25,265	543,002
1975	23,859	3,658	27,517	570,519
1976	30,072	4,627	34,699	605,218
1977	27,274	4,080	31,354	636,572
1978	27,964	3,452	31,416	667,988
1979	28,902	3,381	32,283	700,271
1980	32,177	3,241	36,418	736,671
1981			32,798	769,469

Activities of ABBA Affiliate Associations

Arkansas Brahman Field Day hosts greet ABBA President Frank D. Connell. The Dickie Haynes family of Sheridan, AR hosted this year's Arkansas Brahman Association Field Day and Junior Show at their ranch in Sheridan. Over 200 people attended this year's field day and junior show. Pictured are (left to right): Stephanie, Brenda, Marc and Dickie Haynes with ABBA President Frank D. Connell of Thornton, TX.

Carolinas Brahman Breeders Association held a field day on August 2, 1980. Pictured above are officers elected to the Carolinas Association (left to right): William Roddy, Treasurer; J. Lewis Patterson, Director; Bill Robertson, Vice-president; Elizabeth P. Cook, Secretary; John D. Coble, President and Jerry Carnes, Director.

Activities of ABBA Affiliate Associations

Georgia Brahman Sale committee members met recently to set the date for the Georgia State Brahman sale. Pictured are (left to right): Lionel Chambers of Sealy, Texas representing the sale management firm, with committee members Van Tucker of Tifton, Georgia and Randy Rigsby of Camilla, Georgia. The Georgia Association hosts an annual sale.

The current officers for the Eastern Brahman Association are (left to right): Secretary - Don Berry of Kissimmee, FL; Treasurer - Mike Partin of Kissimmee, FL; President - Wayne Collier of Wauchula, FL; and Vice-President - Greg Carlton of Clewiston, FL. The members of the Eastern Brahman Association recently donated $500 to the junior members to be used as operating capital for the beginning organization.

Activities of ABBA Affiliate Associations

Current officers and directors for the Oklahoma Brahman Breeders Association are (back row, left to right): Gerald Barlow - Director; Charles White - Director; Frank Welch - Secretary-Treasurer; Lonnie Manuel - Director; Dr. Thomas Howard - First Vice-President; Fred E. Butler, Jr. - Second Vice-President. (Front row, left to right): I. P. Renfro - President; Clarence Hardy - Director; Virginia Barlow - Reporter; and Charles Key - Director.

Hosts for the Mississippi Field Day held in 1981 were (left): Dr. and Mrs. James P. Wood and (right) Mr. and Mrs. Marshall Wood of Waynesboro, MS. Marshall Wood is currently serving as a director for the Mississippi Brahman Association.

Activities of ABBA Affiliate Associations

Picutured above are recently elected Texas Brahman Association officers (left to right): President - Henry Clay Koontz of Placedo, TX; Secretary-Treasurer - Margaret Watkins of Ganado, TX; First Vice-President - Jim Brown of Belton, Tx and Second Vice-President - Bruce Lowrie of Ft. Worth, TX.

The Western Brahman Breeders Association recently got together with the Reno Brahman Clinic sponsors John and Vivian Joyce to hold a successful clinic. Pictured are (left to right): Past WBBA President - Chet Edmunds of Riverside, CA; ABBA President Frank Connell; Past ABBA President - Dr. Richard Forgason; Sponsors - John and Vivian Joyce of Wharton, TX; Orben Brackin of Reedley, CA; Charles White of Chowchilla, CA; and Mr. and Mrs. William R. Cowan of Animas, NM.

Women of the ABBA

The Ladies of the ABBA, though only recently organized, have been an important part of the ABBA's work. One of their current projects (1981) has been the raising of $25,000 for ABBA youth activities. Pictured above are the 1981 L'ABBA officers (left to right): Margaret Watkins, Secretary; Betsy Boyt, Social Chairman; Sue Seaberg, President; Monica Long, Vice President and Sugie Sartwelle, Treasurer.

Emily Dial, 3X Ranch, Placedo, Texas has become the first woman to be entered on the American Brahman Breeders Association's List of Official Judges in its 57 year history.

Mrs. Dial comes from a long line of Brahman breeders. One of her ancestors was Captain John N. Keeran. (See Chapter 2, "America's First Indian Cattle.")

Under her guidance cattle with the 3X brand have become a familiar sight in the winner's circle at ABBA shows.

Index

B

F

G

Q

R

COUNTRIES THAT HAVE IMPORTED BRAHMANS FROM THE UNITED STATES (1933-1980)
(Year of Initial Importation)

1933	Australia
	Mexico
1937	Cuba
1938	Colombia
	Costa Rica
	Fiji Islands
	Puerto Rico
	Tahiti
1939	Canada
	Venezuela
1940	Peru
1941	Argentina
	Panama
1944	Hawaii
1945	Brazil
	Guam
	Jamaica
1946	Guatemala
	Haiti
	Surinam
1948	El Salvador
1949	Ecuador
	Honduras
	Mauritius
1950	Dominican Republic
	Nicaragua
1951	Liberia
	Martinique
1952	Cameroon
	Philippines
1953	Guadeloupe
	South West Africa
1954	British Honduras
	Israel
	New Guinea
	Republic of South Africa
	Thailand
1955	Madagascar
	Mozambique
	Rhodesia (Zimbabwe)
1957	Canal Zone
	Chile
	Korea
	Zaire
1958	Bolivia
1960	British Guiana
	Cambodia
	Paraguay
1961	Virgin Islands
1962	Angola
	Antigua
	Dominica
1963	Grand Cayman
	Tortola
	Trinidad
1964	Botswana
	Nigeria
1966	Spain
1969	Zambia
1971	Guyana
1973	Iran
	Malaysia
1974	Taiwan
1975	Swaziland
1977	Abu Dhabi
1979	French Guiana
1980	China

On the Rancho Don Tomas of San Luis de Potosi, Mexico, three young Mexican lads demonstrate the docility of a Brahman bull when properly handled.

Brahman cattle have practically revolutionized the cattle industry in some parts of the world. One of the regions that has benefited the most from their introduction is Central America. These purebred Brahmans are grazing in the noonday sun.

Pictured above is one of the early Lykes Brothers' Brahmans that was exported to Yucatan, Mexico. This bull is *Semental de Rearienta.*

These cattle on the Dadanawa Ranch Rupunui, British Guiana represent a mixture of "common cattle" (principally Criollo), and Brahmans. The infusion of Brahman blood has preserved - and in some cases - made possible the raising of beef and dairy cattle in many parts of the world.

Henry Clay Koontz
Placedo, TX

Torolandia
PANAMA

Bob Parish
Houston, TX

Neil E. Potter
Guatemala City, GUATEMALA

Russell W. Richard
Youngsville, LA

David J. Robichaux, Jr.
Thibodeaux, LA

Urbano Santos
Eagle Pass, TX

C.H. Beville
Bushnell, FL

Dr. & Mrs. Edward J.
LeBlanc, Jr.
Lafayette, LA

NJ Brahman
New Boston, TX

Henry O. Partin & Sons
Kissimmee, FL

Pruett's Red Brahmans
Jasper, AL

Randy Rigsby & Sons
Camilla, GA

E.D. Rogers
Zolfo Springs, FL

Sartwelle Brahman
Ranch, Ltd.
Houston, TX

Marcus Shackelford
Wauchula, FL

Madera Cattle Co.
Houston, TX

J.H. Novak & Sons
Danbury, TX

Johnnie Mack Pierce
Thornton, AR

Johnny D. Reich
Blanket, TX

Raimundo Riojas E.
GUATEMALA

Forbes Ranch
Greenwell Springs, LA

R.B. Sewell
DeRidder, LA

G.P. Smith, Jr.
Longview, TX

Sociedad Anonima
San Luis
Guayaquil, ECUADOR